Systems Biology

PROPERTIES OF RECONSTRUCTED NETWORKS

Bernhard Ø. Palsson

University of California, San Diego

CAMBRIDGE
UNIVERSITY PRESS

CAMBRIDGE UNIVERSITY PRESS
Cambridge, New York, Melbourne, Madrid, Cape Town, Singapore, São Paulo, Delhi

Cambridge University Press
32 Avenue of the Americas, New York, NY 10013-2473, USA

www.cambridge.org
Information on this title: www.cambridge.org/9780521859035

First published 2006
Reprinted 2007

Printed in the United States of America

A catalog record for this publication is available from the British Library.

Library of Congress Cataloging in Publication Data

Palsson, Bernhard.
Systems biology : properties of reconstructed networks / Bernhard O. Palsson.
    p.   cm.
Includes bibliographical references and index.
ISBN-13: 978-0-521-85903-5 (hardback)
ISBN-10: 0-521-85903-4 (hardback)
1. Computational biology.   2. Genomics.   3. Bioinformatics.   I. Title.
QH324.2.P35   2006
572.80285–dc22   2005035008

ISBN   978-0-521-85903-5 hardback

080107-Q8

TO MAHSHID

Contents

Preface

In 1995, the first full genome sequence became available, ushering in the genome era. Since then, a large number of high-throughput technologies have enabled us to define the molecular parts catalogs of cellular components. Although these catalogs are still incomplete, it is now possible to reconstruct, based on this and other information, genome-scale networks of biochemical reactions that take place inside cells. This process of network reconstruction, followed by the synthesis of *in silico* models describing their functionalities, is the essence of systems biology.

The functions of reconstructed networks are defined by the interconnections of their parts. Since these connections involve chemical reactions, they can be described by a stoichiometric relationship. The stoichiometric matrix, which contains all such relationships in a network, is thus a concise mathematical representation of reconstructed networks. This matrix comprise integers that represent time- and condition-invariant properties of a network. It may therefore be expected to represent a key in the study of the functionalities of complex biochemical reaction networks. Its content and associated information effectively constitute a biochemically, genetically, and genomically structured database.

This book is focused on the stoichiometric matrix. In order to satisfactorily understand the material, good knowledge of linear algebra and of biochemistry is needed. Most of the mathematical concepts and principles, when properly interpreted, have a direct biological and chemical meaning. This text thus tries to relate what might be seen as abstract mathematical quantities to real biological and chemical features.

Like it or not, the ability to reconstruct genome-scale reaction networks will firmly thrust biology into the domain of systems science. Not just any systems science, but systems biology and bioengineering, where the underlying biochemical and genetic processes set the stage. Biological education in the future will undoubtedly involve more mathematics, and thus new generations of biologists should be able to readily deal with this material.

This is a personal book. The author has been working on the construction of mathematical descriptions of complicated biochemical reaction networks for more than 20 years. At the inception of this activity, the general view was that such an exercise was purely theoretical and had no real biological relevance. However, with the advent of the genome era, we now have the necessary biological data to build realistic genome-scale networks and relate their properties to observable phenotypes. This field has grown in scope over the last 5 to 10 years and has become quite broad. In a way, this book was written by the author, for the author, in an attempt to organize this field, the concepts it contains, what has been accomplished, and what may potentially lie ahead. Hopefully it will benefit others.

The author has many people to thank for their help in preparing this text. The biggest and most important acknowledgment is to his wife, Mahshid, who has tolerated and accepted many hours of absence during which the contents of this book were conceived, formulated, and brought into practice. Without her patience and support over the past 20 years, this book would not have been possible.

Three individuals influenced my career development and thus contributed to this text being written. As an undergraduate working in the biochemistry laboratory of Sigmundur Gudbjarnason, I learned the wonders of the world of enzymes, their purification, and kinetic characterization. During the analysis of data in his lab, I came to the realization that enzymes are interesting catalysts, but just that. Hundreds if not thousands of them had to come together to reproduce the living process that I was so interested in. I thus selected chemical engineering as a field of study, as it was the only major within which I could study life, chemical, and systems sciences. I was fortunate to join Edwin Lightfoot's laboratory for my PhD studies. Although he had not worked much with systems analysis in molecular biology, he immediately recognized its importance and was willing to support me in my pursuit of such analysis for my PhD studies. Needless to say, systems analysis in biology in the early 1980s was seen as a "dead-end career" and "professional suicide." After joining the University of Michigan as a faculty member I was forced to pursue other, more fundable sources of work. In the early 1990s I became aware of Lee Hood's revolutionary impact on biology through the development of high-throughput approaches. He became an inspiration through his vision and leadership, and eventually a good friend. I am thankful for the positive influence that these three individuals have had on my career development and their impact on the development of the material in this book.

In my transition from the University of Michigan to the University of California, San Diego (UCSD), I had the pleasure of being hosted by Jens Nielsen and John Villadsen at the Technical University of Denmark in the spring of 1996. During this four-month leave, sponsored by the Fulbright and Ib Heinriksen foundations, I pondered the impact that the first full genome-sequences would have, and as a result some of the conceptual foundation for this book was laid down. In 2000, I was appointed as the Hougen visiting professor at the University of Wisconsin. I prepared the Hougen lecture series that fall. These lecture notes proved to be the first outline of the material contained in this book.

The Whitaker Foundation generously supported the preparation of this book through their Teaching Materials Program. It is very hard for professors to prepare textbooks in today's academic environment given the general lack of staff support and the ever-growing demands on faculty members' time. Without the support of the Whitaker Foundation, this book would not have been written.

Marc Abrams managed the technical coordination of this book. He captured most of the material (text, artwork, and references) and translated it into a LaTeX file for publication. He also prepared many of the original illustrations. His patience, persistence, and perseverance during the book-writing process has been much appreciated.

I have been fortunate to have had a series of outstanding students during my stay at UCSD. Special thanks go to many of these excellent students for helping out with several of the chapters in this book: Scott Becker, Chapter 7 and Appendix B; Natalie Duarte, Chapter 3; Iman Famili, Chapters 3, 6, 8, 10, and 14; Adam Feist, Chapter 15 and Appendix B; Markus Herrgard, Chapters 4 and 16; Andrew Joyce, Chapter 4; Jason Papin, Chapters 5, 9, and 13; Nathan Price, Chapters 9 and 14; Jennifer Reed, Chapters 3, 15, 16, and Appendix B; and Ines Thiele, Chapter 13. Thanks also to Henry Kang for editorial assistance.

Four of my other students have been influential in the development of this material: Joanne Savinell, who wrote the first comprehensive PhD thesis on stoichiometric networks in the late 1980s; Amit Varma, who built a biochemically comprehensive *E. coli* model in the early 1990s; Jeremy Edwards, who de facto became the first "high-throughput *in silico* biologist" by building the first genome-scale metabolic models in the late 1990s; and Christophe Schilling, who developed the extreme pathways in the late 1990s and who decided that commercial-strength software and services were needed for genome-scale network reconstruction and model building. I am a cofounder of the resulting Genomatica, Inc. Homework sets and network reconstructions related to this book will be posted on http://systemsbiology.ucsd.edu.

A few books have influenced my thinking over the years. The book by Daniel Atkinson (CELLULAR ENERGY METABOLISM AND ITS REGULATION, 1977) was my first exposure to the analysis of metabolism from a systems and engineering perspective and had a strong influence early on in my career. The book by Jens Reich and Evgeni Selkov (ENERGY METABOLISM OF THE CELL, 1981) had a profound influence on my early thinking. This book is conceptually rich and a pioneering effort toward the quantitative systems analysis of biochemical reaction networks. The recent book of Antoine Danchin (THE DELPHIC BOAT, 2003) is a masterful biological analysis of the contents genomes and what they tell us. The many writings of Ernst Mayr (e.g., THIS IS BIOLOGY, 1997) provide decades worth of the author's perceptive thinking about the basic difference between biology and the physicochemical sciences, a divide that systems biology now tries to bridge.

Many other individuals have directly or indirectly (willingly or unwillingly) influenced the material in this book. They include Adam Arkin, Laszlo Barabasi, Dan Beard, Sydney Brenner, Antony Burgard, George Church, Frank Doyle, John Doyle, David Fell, David Galas, Igor Goryanin, Vassily Hatzimanikatis, David Haussler, Leland Hartwell, Reinhard Heinrich, Jay Keasling, Marc Kirschner, Hioraki Kitano, Stefan Klamt, Choul-Gyun Lee, Sang Yup Lee, William Loomis, Costas Maranas, Harley McAdams, Terry Papoutsakis, Uwe Sauer, Mick Savage, Michael Savageau, Stefan Schuster, Daniel Segré, Lucy Shapiro, Jurg Stelling, Gilbert Strang, Shankar Subramaniam, and Masaru Tomita. Many thanks to these colleagues for the stimulating discussions over the years.

The author hopes that this book will be the beginning of courses and textbooks that will formalize the emerging systems biology paradigm of components to networks to models to computed phenotypes.

La Jolla
May 2005

CHAPTER 1

Introduction

Suddenly, systems biology is everywhere. What is it? How did it arise? The driving force for its growth is high-throughput (HT) technologies that allow us to enumerate biological components on a large scale. The delineation of the chemical interactions of these components gives rise to reconstructed biochemical reaction networks that underlie various cellular functions. Systems biology is thus not necessarily focused on the components themselves, but on the nature of the links that connect them and the functional states of the networks that result from the assembly of all such links. The stoichiometric matrix represents such links mathematically based on the underlying chemistry, and the properties of this matrix are key to determining the functional states of the biochemical reaction networks that it represents.

1.1 The Need for Systems Analysis in Biology

Biological parts lists

During the latter half of the 20th century, biology was strongly influenced by reductionist approaches that focused on the generation of information about individual cellular components, their chemical composition, and often their biological functions. Over the past decade, this process has been greatly accelerated with the emergence of genomics. We now have entire DNA sequences for a growing number of organisms, and we are continually delineating their gene portfolios. Although functional assignment to these genes is presently incomplete, we can expect that we will eventually have assigned and verified function for the majority of genes on selected genomes. Extrapolation between genomes will then most likely accelerate the definition of what amounts to a "catalog" of cellular components in a large number of organisms. Expression array and proteomic technologies

Reductionist approach:

Components
biology

HT analytical chemistry:
genomics,
transcriptomics,
proteomics

20th-Century biology

Integrative approach:

Systems
biology

Integrative analysis:
bioinformatics,
mathematical models,
computer (*in silico*)
simulation

21st-Century biology

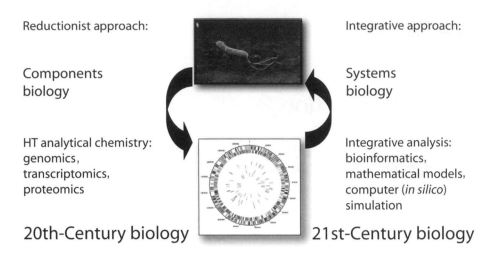

Figure 1.1: Illustration of a paradigm shift in cell and molecular biology from component to systems analysis. Redrawn from [152].

give us the capability to determine when a cell uses particular genes, and when it does not (left side in Figure 1.1). At the beginning of the 21st century, this process was unfolding at a rapid rate, driving a fundamental paradigm shift in biology.

Beyond bioinformatics

The advent of high-throughput experimental technologies is forcing biologists to view cells as systems, rather than focusing their attention on individual cellular components. Not only are high-throughput technologies forcing the systems point of view, but they also enable us to study cells as systems. What do we do with this developing list of cellular components and their properties? As informative as they are, these lists only give us basic information about the molecules that make up cells, their individual chemical properties, and when cells choose to use their components.

How do we now arrive at the biological properties and behaviors that arise from these detailed lists of chemical components? It is now generally accepted that the integrative analysis of the function of multiple gene products has become a critical issue for the future development of biology. Such integrative analysis will rely on bioinformatics and methods for systems analysis (right side of Figure 1.1). It is thus likely that over the coming years and decades biological sciences will be increasingly focused on the systems properties of cellular and tissue functions. These are the properties that arise from the whole and represent biological properties. These properties are sometimes referred to as "emergent" properties since they emerge from the whole and are not properties of the individual parts.

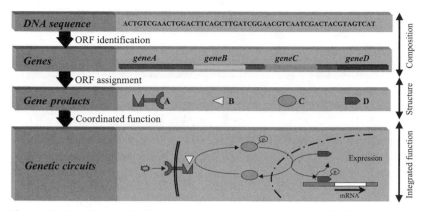

Figure 1.2: Genetic circuits. From sequence, to genes, to gene product function, to multicomponent cellular functions. Prepared by Christophe Schilling.

Genetic circuits

The relationship between genetics and cellular functions is hierarchical and involves many layers, some of which are illustrated in Figure 1.2. Gene sequences allow for the identification of open reading frames (ORFs). The base pair sequence of the ORFs in turn allows for the functional assignment of the defined gene. Although not always unambiguous, such assignments are being carried out with increasing accuracy, due to our expanding biological databases. Sequence is important and so is the functional assignment of ORFs. However, the interrelatedness of the genes may prove to be even more important. Establishing these relations and studying their systemic characteristics is now necessary.

Cellular functions rely on the coordinated action of the products from multiple genes. Such coordinated function of multiple gene products can be viewed as a "genetic circuit" (some synonyms that are commonly used are "cellular wiring diagrams" and modules). The term genetic circuit is used here to designate a collection of different gene products that together are required to execute a particular cellular function. The functions of such genetic circuits are diverse, including DNA replication, translation, the conversion of glucose to pyruvate, laying down the basic body plan of multicellular organisms, and cell motion. It is likely that we will view cellular functions within this framework and the physiological functions of cells and organisms as the coordinated or integrated functions of multiple genetic circuits. Consequently, we will need to develop the conceptual framework within which to describe and analyze these circuits.

Not all the properties of genetic circuits are clear at present, but some important ones are summarized in Table 1.1. For many of these characteristics, it is also clear what methodology is needed to describe and analyze

Table 1.1: Some of the characteristics of genetic circuits and the analysis methods required.

Characteristic	Analysis method
They are complex	Bioinformatics
They are autonomous	Control theory
They are robust	System science
They function to execute a physicochemical process	Transport and kinetic theory
They have "creative functions"	Bifurcation analysis
They are conserved, but can adjust	Evolutionary dynamics

them. Genetic circuits tend to have many components; they are complex. From the standpoint of system science, they are "robust," i.e., in many, but not all cases, one can remove their components without compromising their overall function.

Accepting the concept of a gene circuit seems straightforward. However, the implications of this acceptance are quite profound. We will view bioinformatics as a way to establish, classify, and cross-species correlate genetic circuits. The beginning of such classification is illustrated in Figure 1.3. Metabolism, information processing, and cellular fate processes represent some of the major categories of genetic circuits. Considerable unity in biology is likely to result in conceptualizing biological functions as genetic circuits. From this standpoint, gene therapy may no longer be viewed as replacing a "bad" gene, but instead fixing a "malfunctioning" genetic circuit. Evolution may be viewed as the "tuning" or "honing" of circuits to improve performance and chances of survival. Classifying organisms based on the types of genetic circuits they possess may lead to "genomic taxonomy." *Ex vivo* "evolutionary" procedures for designing genetic circuit performance are emerging [99, 258]. Understanding the function of genetic circuits will become fundamental to applied biology, in fields as diverse as metabolic engineering and tissue engineering.

The concept of a genetic circuit as a multicomponent functional entity (either in time or space, or both) is an important paradigm in systems biology. It will be a fundamental component in our treatment of the relationship between genetics and physiology. the genotype–phenotype relationship. Individual genetic circuits do not operate in isolation, but in the context of other genetic circuits. The assembly of all such circuits found on a genome produces cellular and organismic functions and leads to hierarchical decomposition of complex cellular functions. Thus, the need for genome-scale analysis arises. This need in turn leads to viewing the genome as the "system."

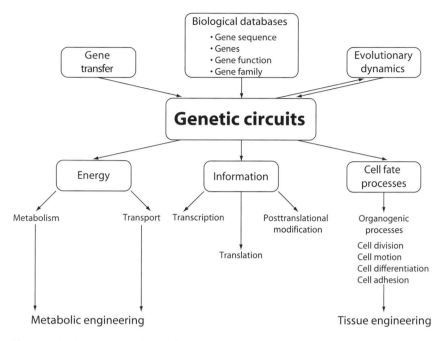

Figure 1.3: Coarse-grained classification of the types of genetic circuits that are found on genomes. Some major categories are indicated, in particular to indicate that some underlie the important metabolic and tissue engineering applications of cell and molecular biology. Prepared by Christophe Schilling.

1.2 The Systems Biology Paradigm

The ability to generate detailed lists of biological components, determine their interactions, and generate genome-wide data sets has led to the emergence of systems biology [101]. The process comprises four principal steps (Figure 1.4). First, the list of biological components that participate in the process of interest is enumerated. Second, the interactions between these components are studied and the "wiring diagrams" of genetic circuits are constructed. This process is one of biochemical reaction network reconstruction and is covered in detail in Part I of this text. Third, reconstructed network are described mathematically and their properties analyzed (Part II). Computer models are then generated to analyze, interpret, and predict the biological functions that can arise from the reconstructed networks (Part III). Fourth, the models are used to analyze, interpret, and predict experimental outcomes. Prediction essentially corresponds to generating specific hypotheses that can then be experimentally tested. These *in silico* models of reconstructed networks are then improved in an iterative fashion [152].

There is much creative work that has led to the development of high-throughput technologies (step 1). Many different mathematical methods

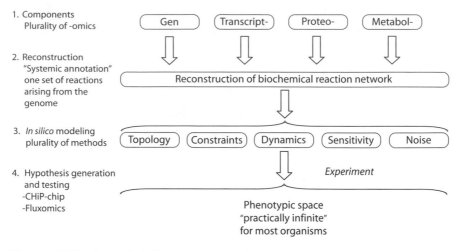

Figure 1.4: The four principal steps in the implementation of systems biology. Note that the second step is unique, while the others are diverse, and is the interface between high-throughput data and *in silico* analysis.

have been formulated for the analysis of biochemical reaction networks (step 3), and the phenotypic space explored by experimentation (step 4) is essentially infinite. In contrast, the reconstruction effort leads to one result. The reconstruction should culminate in the generation of the set of chemical reactions that take place inside a cell and underlie its function. Although systems biology is currently thought of as a cell-scale effort representing a genome-enabled science, it is likely that it will be conceptualized as a broader field as it develops. We will begin to talk about the systems biology of tissues, through networks of cellular interactions, and so on.

Systemic annotation

The unity represented by step 2 in Figure 1.4 leads to an effort to create a *two-dimensional* annotation of a genome (Figure 1.5). The classical component annotation of a genome leads to the identification of open reading frames, their location, and often the corresponding DNA regulatory sequences, a one-dimensional list of components. The open reading frames can then be assigned function based on homology searches of known genes. The two-dimensional annotation accounts for not only the components, but all their chemical states (represented as rows in the table in Figure 1.5) and the links between them. The latter are the columns in the table and ideally should represent the stoichiometric coefficients that correspond to the underlying chemical transformations that are possible between the components. This table represents the full genome-scale stoichiometric matrix for a genome.

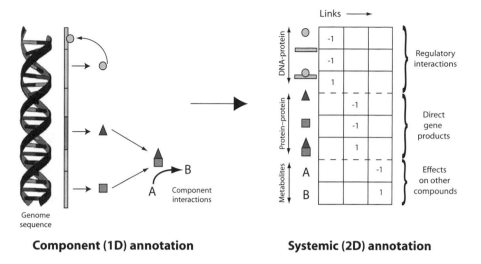

Component (1D) annotation   **Systemic (2D) annotation**

Figure 1.5: Systemic or two-dimensional annotation of genomes: the origin of the stoichiometric matrix. From [155].

Calling for the formulation of this matrix may represent as bold a statement as asking for the full base pair sequence of the human genome some 20 years ago. However, progress is being made. Genome-scale metabolic networks have been reconstructed for microorganisms. We are in the process of beginning to define signaling networks and transcriptional regulatory networks. Sometimes events in such networks are known chemically, but sometimes we only have causal relationships, which eventually will be converted into chemical equations once the underlying mechanisms are discovered.

Hierarchical thinking in systems biology

We are quite used to thinking hierarchically about DNA. We think about a base pair as the irreducible unit of DNA sequence. Then we talk about codons, introns, exons, alleles, and chromosomes, and other measures of DNA size. We will need to adapt similar hierarchical thinking about the genome-scale stoichiometric matrix. The irreducible elements in a network are the elementary chemical reactions. These can combine into reaction mechanisms, many reactions into modules or motifs, pathways can form, and sectors can be defined. Currently, such coarse graining of a network often relies on somewhat ill-defined notions of hierarchical structure.

Our understanding of how to hierarchically decompose a network is likely to improve as we begin to build genome-scale networks and are able to define their properties. Components that always function together in steady or dynamic states normally would fall into modules. Correlated

Biology root

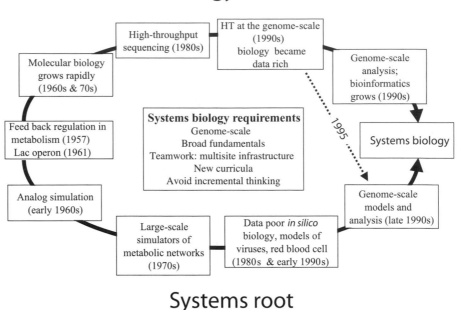

Figure 1.6: The two roots of systems biology.

subsets of reactions do appear in the delineation of steady-state properties of networks ([163], Chapter 9). Time-scale separation is often used for temporal decomposition of complex systems, and the stoichiometric matrix does seem to play a role in this formation of dynamics pools [109, 149] that represent dynamic course graining of a network.

Historical roots

Although it is often stated that reductionist thinking has characterized molecular biology, it does not mean that integrative thinking has not taken place. The first genetic circuits were indeed mapped out more than 40 years ago (Figure 1.6). The feedback inhibition of amino acid biosynthetic pathways was discovered in 1957 [225, 257], and the transcriptional regulation associated with the glucose–lactose diauxic shift led to the definition of the *lac* operon [12, 124]. These regulatory mechanisms began the unraveling of the molecular logic that underlies cellular processes.

In the decades following these discoveries, molecular biology blossomed as a field. In the 1980s we began to see the scale-up of some of the fundamental experimental approaches of molecular biology. Automated DNA sequencers began to appear and reached genome-scale sequencing in the mid-1990s. Automation, miniaturization, and multiplexing of various assays led to the generation of additional "omics" data types. The large volumes of data generated by these approaches led to a rapid growth of the field of bioinformatics. Although this effort was mostly focused on statistical models and object classification approaches in the late 1990s, it became recognized that a more formal and mechanistic framework was needed to systematically analyze multiple high-throughput data types [153]. This need led to calls for genome-scale model building.

Following the events of the late 1950s and early 1960s, efforts were initiated to formulate mathematical models to simulate the functions of the newly discovered genetic circuits. Even in the early 1960s, before digital computers became available, the function of such circuits was simulated on analog computers [78]. These efforts grew to the dynamic simulation of large metabolic networks in the 1970s [69, 123, 253, 255]. By the late 1980s and early 1990s, cell-scale models of the human red cell had appeared [106], genome-scale models of viruses were formulated, and large-scale models of mitosis appeared [146]. The advent of genome-scale sequencing led to the first genome-scale metabolic models for bacteria [50, 51].

These two roots of systems biology are illustrated in Figure 1.6. The upper branch had much greater presence in the scientific community, dazzling us with a never-ending stream of discoveries and exciting technologies. One might say that this was the "biology" root to systems biology. The lower branch never gained much notoriety, although, unlike in the United States, this activity was reasonably prominent in Europe. Systems modeling and simulation in molecular biology was seen as purely theoretical and not a contributor to understanding real biology. However, with biology now having become a "data-rich" field, the need for theory, model building, and simulation has emerged. One might think of this branch as the "systems" root to the emergence of systems biology.

These two branches must now merge to further the field. While there are many books and sources on the "biological root," few exist for the "systems root." This book is an attempt to meet this need, although real genome-scale analysis is still the material of cutting-edge research. Thus, this initial effort is conceptual and illustrative in nature with references to the genome-scale studies that have appeared.

1.3 About This Book

Purpose

The availability of annotated genome sequences in the mid to late 1990s enabled the reconstruction of genome-scale metabolic networks [37]. Similar reconstructions of signaling and transcriptional regulatory networks are now beginning to appear [85, 213]. The topological structure and functional properties of these networks can now be studied. More importantly, for the first time, we can analyze, interpret, and predict the phenotypic functions that such networks could produce. The stoichiometric matrix is a compact mathematical representation of biochemical reaction networks. It represents the interface between the HT data world and that of *in silico* analysis, e.g., Figure 1.4, and the two-dimensional annotation of a genome. The purpose of this book is to describe how the stoichiometric matrix is formed, what its basic properties are, and how it can be used to analyze the functional states of networks.

Approach

We will first outline some of the basic concepts of systems biology in Chapter 2. We will then divide the material into three parts.

> **Part I** will briefly review three types of networks – metabolic, regulatory, and signaling – and show how they are comprised of underlying biochemical reactions. The efforts to reconstruct them are intensive in analyzing the data from various HT experimental technologies and legacy (bibliouric) data. Reconstructions basically culminate in the formation of a chemically, genetically, and genomically (BIGG) structured database that represents all the data types simultaneously. Once curated, a genome-scale reconstruction represents a BIGG structured integration of the available information about a cell or an organism.

> **Part II** will describe the formation of the stoichiometric matrix, $\mathbf{S}$, including its function as a mathematical mapping operation, the chemical constraints on its structure, and its topological properties. An understanding of basic linear algebra now becomes essential to the reader. The topological properties of the stoichiometric matrix are then outlined and methods for their analysis described. We then explore the more subtle and intricate properties of the stoichiometric matrix. To do so, we need to study its fundamental spaces associated with $\mathbf{S}$ and will thus require an intermediate-level understanding of linear algebra. The two null spaces of $\mathbf{S}$ contain systemically defined reaction pathways and concentration conservation quantities. The row and column spaces of $\mathbf{S}$ contain the dynamic flux vector and the time derivatives, respectively. These two spaces are thus key to studying the transient function and the underlying thermodynamics. The transition from Part I to Part II may be challenging for some life scientists, but it is important for mastering systems biology.

> **Part III** will describe the mathematical methods that have been developed to interrogate the properties of reconstructed networks. The reconstructions and their associated information are not sufficient to completely define the state of a network. Flexibility in function exists, leading to constraint-based analysis. This approach is consistent with the

biological reality of operating under governing constraints, but allowing for evolution within them to adapt and improve biological function.

1.4 Summary

➤ Detailed biological parts catalogs of cells have emerged.

➤ The chemical and causal interactions of these parts are being documented.

➤ Cellular "wiring diagrams" representing genetic circuits and genome-scale networks are being reconstructed.

➤ The systems biology paradigm of "components→ networks→ *in silico* models→ phenotype" has arisen.

➤ Two-dimensional or systemic annotation of genomes is emerging and represents unity of effort in systems biology through network reconstruction.

➤ Network reconstruction is described by a BIGG structured data base.

➤ The stoichiometric matrix describes the reconstruction mathematically and thus it becomes a key to the field of systems biology.

➤ Systems biology is inherently mathematical.

1.5 Further Reading

Aebersold, R., Hood, L.E., and Watts, J.D., "Equipping scientists for the new biology," *Nature Biotechnology*, **18**:359 (2000).

Ge, H., Walhout, J.M., and Vidal M., "Integrating "omic" information: A bridge between genomics and systems biology," *Trends in Genetics*, **19**:551–560 (2003).

Hasty, J., McMillen, D., Isaacs, F., et al., "Computational studies of gene regulatory networks: In numero molecular biology," *Nature Reviews Genetics*, **2**:268–279 (2001).

Holland, J.H., EMERGENCE, Addison-Wesley, New York (1988).

Ideker, T., Galitski, T., and Hood, L., "A new approach to decoding life: Systems biology," *Annual Review of Genomics and Human Genetics*, **2**:343–372 (2001).

Kanehisa, M., POST-GENOME INFORMATICS, Oxford University Press, New York (2000).

Palsson, B.O., "What lies beyond bioinformatics?" *Nature Biotechnology*, **15**:3–4 (1997).

CHAPTER 2

Basic Concepts in Systems Biology

In the early 1960s, there was a bifurcation of emphasis in biology. Molecular biology had arrived, providing a growing understanding of DNA, protein, and other chemical components of cells. A science was emerging that had rigor in terms of analytical chemistry and controlled experimentation, and relevance to biochemical and genetic functions of cells and occasionally to their phenotypes. Holistic emphasis in biology, which had primarily been practiced through physiology, faded into the background as it is much more difficult to state hypotheses, do controlled experiments, or execute the scientific process for the behavior of systems and networks in biology. However, as outlined in the introductory chapter, this situation has now changed. We now have technology that allows for the detailed enumeration of biological components, enabling us to study cells and complex biological processes as systems. As a consequence, systems biology has arisen as a new field. This new field does not yet have a well-defined and articulated conceptual basis. In this chapter, we will attempt to collect some of the key issues that represent to the conceptual foundations of systems biology. Its content is not intended to be, and cannot be, complete but rather represents an attempt to initiate this process.

2.1 Components vs. Systems

Biological components all have a finite turnover time. Most metabolites turn over within a minute in a cell, mRNA molecules typically have 2-hour halflives in human cells [256], 3% of the extracellular matrix in cardiac muscle is turned over daily, and so forth. So a cell that you observe today, compared with the same cell yesterday, may only contain a small fraction of the same molecules. Similarly, cells have finite lifetimes. The cellularity of the human bone marrow turns over every 2–3 days, the renewal rate of

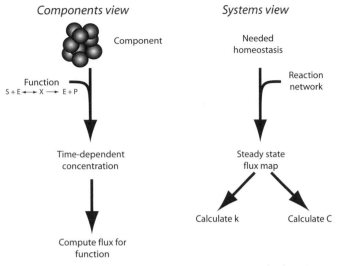

Figure 2.1: A contrast between the components view (left) and the systems view (right).

skin is of the order of 5 days to a couple of weeks, the lining of the gut epithelium has a turnover time of about 5–7 days, and slower tissues like the liver turn over their cellularity approximately once a year. Therefore, most of the cells that are contained in an individual today were not there a few years ago. However, we consider the individual to be the same, albeit older. Likewise, we consider one cell to be the same a week later, even if most of its chemical components have turned over. Components come and go, therefore a key feature of living systems is how their components are connected together. The interconnections between cells and cellular components define the essence of a living process.

The difference between the components view of life is different from the systems view in many subtle ways. Here, we try to illustrate this difference by just one example (see Figure 2.1).

- On the left side of Figure 2.1 we see the components point of view of the function of a gene product. When we are looking at one gene product, in this case an enzyme carrying out its function, we study this component by placing it in a beaker with its substrates and then observe the time-dependent disappearance of a substrate and the appearance of a product. The component that we are studying is the centerpiece of this experiment, and it is responsible for concentration changing in a time-dependent manner.
- The right side illustrates a systems viewpoint of a biochemical network. It is not so much the components themselves and their state that matters, contrary to the components view, but it is the state of the whole system that counts. Any biological network will have a nominal

state, which we recognize as a homeostatic state. Thus, the fluxes that reflect the interactions among the components to form the state of the network are dominant variables, and the concentrations of the individual components are "subordinate quantities." The concentrations of the network components are determined first by the flux map, or the state of the network, and then by the kinetic properties of the links in the network.

2.2 Links and Functional States

Two key issues arise from the earlier considerations. The first deals with the nature of the links between components in a biological network, and the second deals with the functional states and the properties of a network that a set of links form.

Links

Links between molecular components are basically given by chemical reactions or associations between chemical components. These links are therefore characterized and constrained by basic chemical rules. In tissue biology, the nature of links between cells is more complicated and often related to higher-order chemistry. We note that a T-cell receptor, for instance, forms a complicated structure in the membrane of a cell and the properties of that structure, and how compatible it is with the complimentary features of another cell determines whether there is communication or links between these cells. Since we are focused on the characteristics of biochemical networks, we will further discuss the chemical nature of links in molecular biology.

The prototypical transformations in living systems at the molecular level are bilinear. This association involves two compounds coming together to either be chemically transformed through the breakage and formation of covalent bonds, as is typical of metabolic reactions or macromolecular synthesis,

$$X + Y \rightleftharpoons X - Y \quad \text{covalent bonds}$$

or two molecules associated together to form a complex that may be held together by hydrogen bonds and/or other physical association forces to form a complex that has a different functionality than individual components,

$$X + Y \rightleftharpoons X : Y \quad \text{association of molecules}$$

Such association, for instance, could designate the binding of a transcription factor to DNA to form an activated site to which an activated polymerase binds. Such bilinear association between two molecules might also involve the binding of an allosteric regulator to an allosteric enzyme that induces a conformational change in the enzyme.

Chemical transformations have certain key properties:

1. *Stoichiometry.* The stoichiometry of chemical reactions is fixed and is described by integral numbers counting the molecules that react and that form as a consequence of the chemical reaction. Thus, stoichiometry basically represents "digital information." Chemical transformations are constrained by elemental and charge balancing, as well as other features. Stoichiometry is invariant between organisms for the same reactions and does not change with pressure, temperature, or other conditions. Stoichiometry gives the primary topological properties of a biochemical reaction network.

2. *Relative rates.* All reactions inside a cell are governed by thermodynamics. The relative rate of reactions, forward and reverse, is therefore fixed by basic thermodynamic properties. Unlike stoichiometry, thermodynamic properties do change with physicochemical conditions such as pressure and temperature. The thermodynamic properties of associations between macromolecules can be changed by altering the sequence of a protein or the base-pair sequence of a DNA binding site. The thermodynamics of transformation between small molecules in cells are fixed but condition dependent.

3. *Absolute rates.* In contrast to stoichiometry and thermodynamics, the absolute rates of chemical reactions inside cells are highly manipulable. Highly evolved enzymes are very specific in catalyzing particular chemical transformations. Cells can thus extensively manipulate the rates of reactions through changes in their DNA sequence.

Therefore, links cannot just form between any two cellular components. The links that are formed are constrained by the nature of covalent bonds that are possible and by the thermodynamic nature of interacting macromolecular surfaces. All of these are subject to the basic rules of chemistry and thermodynamics. The absolute rates are key biological design variables because they can evolve from a very low rate determined by the mass action kinetics based on collision frequencies to a very high and specific reaction rate determined by appropriately evolved enzyme properties. Enzymes evolve to bring molecules into particular orientation to control the rate of appropriately oriented collisions between two molecules that lead to a chemical reaction (see Figure 2.2).

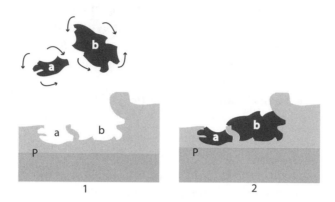

Figure 2.2: A schematic of how the binding sites of two molecules on an enzyme bring them together to collide at an optimal angle to produce a reaction. Panel 1: Two molecules can collide at at random and various angles in free solution. Only a fraction of the collisions lead to a chemical reaction. Panel 2: Two molecules bound at the surface of a reaction can only collide at a highly restricted angle, substantially enhancing the probability of a chemical reaction between the two compounds. Redrawn from [122].

Functional states

Once all the links in a network have been identified and described, its functional states can be determined. We can study the topological properties of a network, but these properties give is only limited information about the actual functional state of a network. The functional states of biological reaction networks are constrained by the physicochemical nature of the intracellular environment (see Figure 2.3). There is a highly developed spatiotemporal organization that orients the biological components and determines the transient nature of the interactions. Interestingly, cells are in a near crystalline state. The protein density in cytoplasm and mitochondria is very close to the protein density in a protein crystal. There are some other notable higher-order properties of biological networks, which will not be detailed here, which include *self-assembly* of components to spontaneously form a functioning network, the *selection* that seems to be at work at all levels in biology, and the interesting notion of *a self* in biology, namely, is a component a part of a network or not?

2.3   Links to Networks

Chemical reactions link components together to form a network. Although we can specify the chemical properties of links in biological networks, it is the way in which a multitude of such links form networks that determines phenotypic functions.

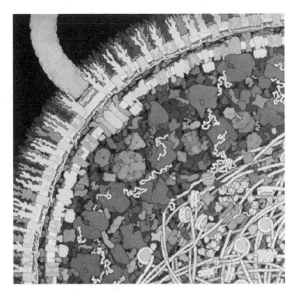

Figure 2.3: The crowded state of the intracellular environment. Some of the physical characteristics are viscosity $> 100 \times \mu_{H_2O}$, osmotic pressure < 150 atm, electrical gradients $\approx 300,000$ V/cm, and a near crystalline state. Copyright David S. Goodsell 1999.

Most biochemical reactions are bilinear. Bilinearity gives the networks a hypergraph property that is topologically nonlinear. The biochemical consequence of this is that biochemical reaction networks form a *tangle of cycles* [186] where different chemical properties and moieties are being transferred throughout the network from one carrier to the next. Perhaps the most familiar of such transformations is the movement of high-energy phosphate bonds between metabolites and proteins. ATP is the primary carrier of such high-energy bonds, and, for instance, a phosphate group is tied to glucose to form glucose-6-phosphate as the first step in glycolysis. The same feature is found in signaling networks whose components are in phosphorylated or dephosphorylated states. Other properties being transferred between molecules are redox potential, 1 carbon units, 2 carbon units, ammonia groups, and so on. This makes biochemical reaction networks highly interwoven.

One interesting feature of biochemical networks as they grow in size is the fact that because of combinatorics, the number of possible functional states that they can take can grow faster than the number of components in a network. This proliferation in the number of functional states seems to occur past some (a relatively low number) components that come together to form a network. Therefore, the number of phenotypic functions derivable from a genome does not linearly scale with the gene number contained in that genome. For instance, the human genome may have only 50% more genes than the genome of *Caenorhabditis elegans*, a small worm, but nevertheless human beings display much more complicated phenotypes and in greater variety. Thus, in general, it is hard to correlate organism complexity and functions to the number of genes that the organism's genome contains.

The fundamental property of biochemical networks of having many possible functional states leads to the possibility of having the same network carry out many functions and displaying many different phenotypic behaviors. An organism does not fully exploit or use all such possible functional states. Many possible states will be useless to the organism in its struggle for survival. Therefore, a limited subset of these functional states needs to be selected and expressed by cells. We are becoming increasingly familiar with the regulatory mechanisms that carry out the selection of functional states. We are unraveling the very complicated transcriptional regulatory networks in single-celled organisms and the signaling networks that coordinate the function of multicellular organisms. As we will discuss in Chapter 16, complex biological reaction networks will have *equivalent* functional states; that is, there are identical overall functional states that differ in the ways in which they use the underlying links in the network.

Some of the key features of biological networks that distinguish them from other networks need to be accounted for in the analysis of their systemic properties. The first basic feature of biological networks is that they evolve; they change with time. They are *time variant*. Principally, such changes occur through the kinetic properties of the links in the network and the changing of the available or active links in the network at any given point in time. The number of available links can be manipulated by regulation of gene expression, by horizontal gene transfer, and by other mechanisms. The second feature that has to be taken into account is the fact that they have a sense of *purpose*. The fundamental purpose is survival. However, in complicated organisms that fundamentally comprise many networks, some will have goals that are subtasks to the overall goal of survival. For instance, the goal of adipocytes would be to collect and store fat if, in their environment, there is an abundance of energy resources. The goal of the mitochondrion, being the powerhouse of the cell, seems to be to maximize ATP production from available resources. Therefore, the study of *objectives*, that is, purpose, of biochemical reaction networks becomes a relevant and perhaps a central issue.

Thus, linking many biological components together forms a network. This network can have many functional states from which a subset is selected. Links, network topology, and the sense of purpose can all change with time or environmental conditions. It is important to be cognizant of the fact that biochemical reaction networks have to operate in the crowded interior of a cell (see Figure 2.3). Thus, the network view of the biological process has to be considered in the context of the three-dimensional

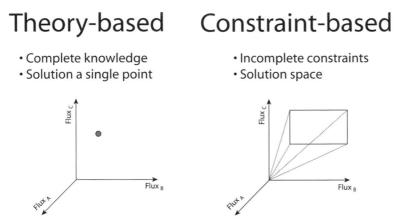

Figure 2.4: Theory-vs. constraint-based analysis. Illustration of finding an exact solution (a point) versus finding a range of allowable solutions (a solution space).

physical arrangement of such networks. These considerations may limit the usefulness of analogies with other man-made networks, such as electrical circuits.

2.4 Constraining Allowable Functional States

The earlier considerations of the nature of links, how they form networks, and how networks form functional states make it likely that *in silico* modeling and simulation of genome-scale biological systems is going to be different than that practiced in the physicochemical sciences. First is the notion that a network can fundamentally have many different states or many different solutions. Which states (or solutions) are picked is up to the cell, and such choices can change over time based on the selection pressure experienced. This difference from the physico-chemical sciences is illustrated in Figure 2.4. All theory-based considerations in engineering and physics lead one to attempt to seek an "exact" solution, typically computed based on the laws of physics and chemistry. However, in biology it appears that not only can a network have many different behaviors that are picked based on the evolutionary history of the organism, but also, as we shall see, these networks can carry out the same function in many different and equivalent ways. Therefore, what are called *silent phenotypes* in biology may be mathematically synonymous to multiple *equivalent network states*. This further leads to an interesting distinction in mathematical modeling philosophy between the key disciplines (Table 2.1).

In physics, the emphasis has always been on deriving theory. Quantum mechanics developed about 100 years ago. Boltzmann derived his famous

Table 2.1: Disciplinary differences in modeling philosophy.

	Equations	Boundary conditions	Nature of solutions
Physics	+++	+	Unique
Engineering	++	++	Design
Biology	+	+++	Multiple and changing

equation prior to that. Theory, as expressed by mathematical equations representing our understanding of fundamental physical mechanisms, has been and probably will continue to be, central to physics. If one wants to then obtain particular solutions to these equations, one has to impose boundary conditions that typically lead to a calculation of a unique solution.

Engineering takes a bit of a departure from this philosophy. The equations used in engineering do not need to be mechanistically correct, in a fundamental theoretic sense, as long as they phenomenologically describe the phenomena and process at hand. Furthermore, the boundary conditions that need to be stated are very important and are often very specific to what an engineer is designing. In engineering, though, one is used to the fact that a problem can have multiple solutions, and this often comes down to the use of design variables to try to optimize a design.

In biology, based on the earlier consideration, we find that the equations needed to describe the physics of the intracellular environment may never be well known, and furthermore, network functionalities evolve and change over time. Therefore, the fundamental equations describing biological functions may be hard to formulate and fully define. On the other hand, the boundary conditions or the constraints under which cells operate and evolve against are easier to identify, state, and use. Constraint-based approaches to the analysis of complex biological systems have proven to be very useful (see Part III of this book).

The constraints under which a cell operates

Cells operate under myriad constraints. There are different ways to classify these constraints, and many authors have discussed them from different points of view. A few will be mentioned here.

- A statement of two very general categories of constraints imposed by natural selection have been described by F. Jacob [102]. They basically are (i) the requirement for reproduction and the genetic mechanisms required to produce offsprings with nonidentical genetic composition of the parent(s), and (ii) the permanent interaction with the environment that imposes thermodynamic constraints of constant flux of matter, energy, and information. The latter constraints are easier to describe in the language of the basic physical laws.

- Danchin [41] in his insightful book about genomes divides the cellular processes and their associated constraints into four general categories: (i) compartmentalization to segregate function in space and to differentiate the "inside" from the "outside"; (ii) metabolism that determines the flow of matter, energy, and redox potential within cells, and its relationship with the outside world; (iii) the transfer of memory to physicochemical processes (i.e., "actuating" inherited information); and (iv) memory transmitted from one generation to the next.

- The author and his collaborators have defined four categories of constraints that can be used to analyze the capabilities of reconstructed biochemical reaction networks [176]: (i) physicochemical constraints, (ii) spatial and topological constraints, (iii) environmental constraints, and (iv) regulatory constraints. These constraints can be mathematically described and used to assess the capabilities of networks (see Chapter 12.).

Viewing regulation as self-imposed constraints, or perhaps as *restraints*, justifies a few more observations in the context of natural selection and organism survival.

Picking candidate states

Cells are subject to inviolable constraints such as those associated with mass and energy balances. Their underlying biochemical networks must obey these and other spatial constraints. These constraints have been called *hard constraints* and, as illustrated by the pentagon in Figure 2.5, give a range of all allowable states of the network. One or more states may be deemed suitable by the cell on the basis of its evolutionary history and current challenges (i.e., the prevailing environmental constraints). A way to exclude all the unwanted states (i.e., those that are unsuitable, or selected against) is to implement a regulatory network that eliminates a large portion of the solution space (the pentagon) and by default forces the expression of the "desired" phenotype. These issues are discussed further in Chapters 12 and 13.

If a state or phenotype is not the best one under given conditions, the solution can move within the allowable range. This change in the selection of a functional state can be accomplished by regulating the expression of the genes that are present at a given point in time and/or by regulating the activity of the corresponding gene products. Such regulation has a relatively short time profile. Over longer times, of course, the components

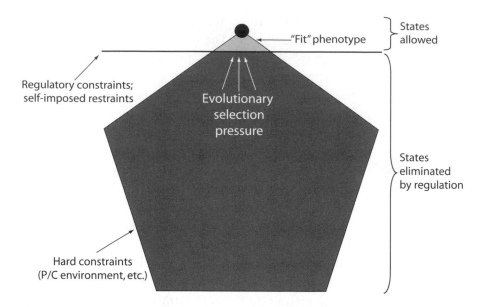

Figure 2.5: Illustration of the constraints on network functions. The pentagon illustrates the range of allowable functions based on hard physicochemical and environmental constraints. The solid line illustrates self-imposed constraints (restraints) produced by regulatory networks; that is, all the states below the line are ruled out by regulatory mechanisms. The dot denotes the desired functional state, which is found among the admissible states after regulatory constraints have been imposed.

of the network can evolve and the properties change slightly, allowing a drift in the phenotypic function of the cell.

Hierarchical organization in biology

Many facets of cellular function and properties are organized hierarchically. The spatial organization of the DNA is shown in Figure 2.6A. The linear dimension of the *E. coli* genome is about 1 mm, while the length of the cell is of the order of 1 μm, a 1000-fold difference. The bacterial genome is thus "folded" a thousand times, in a hierarchically organized fashion. Biochemical reaction networks can be similarly decomposed (Figure 2.6B). Reactions group together into coordinated units that may be colocalized in space, or even compartmentalized. Many such coordinated units can form a larger organized unit, and so forth.

The constraints that apply to the lower levels of organization by necessity will constrain the subsequent higher level functions. This upward application of constraints necessitates a *bottom-up* approach to the analysis of complex biological phenomena. Gödel's completeness theorem in mathematics that showed an axiomatic approach to proving mathematical theorems could not prove that all properties of a system may in a general sense apply to biology. If so, we cannot construct all higher level functions

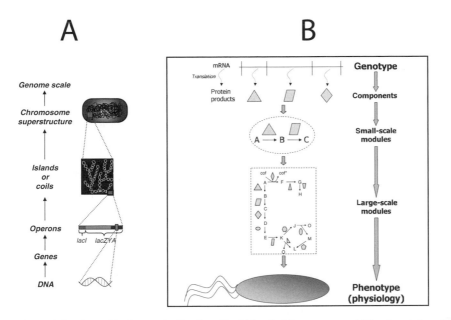

Figure 2.6: Hierarchical organization in cells: (a) bacterial genome and (b) network topology and function. Prepared by Timothy Allen (A) and Jason Papin (B).

from the elementary operations alone. Thus, observations and analyses of system level functions will be needed to complement the bottom-up approach. Therefore, bottom-up and top-down approaches are complementary to the analysis of the hierarchical nature of complex biological phenomena.

The successive adoption of cellular functions over evolution are illustrated in Figure 2.7. The basic biochemistry of cellular processes and the maintenance and expression of the information on the DNA molecule evolved early. This basic set of processes is still found in most organisms today. The genetic code is essentially universal and most proteins are made up of about 20 amino acids. These are basic constraints under which all subsequent cellular processes must operate. The genetic code cannot be predicted from basic theory or physics [39] but is consistent with the basic laws of physics and chemistry. Once picked, it is essentially fixed over evolution. Similarly, most modern proteins are made up of a limited number of motifs, and the basic circuits that lay out the body plan are remarkably conserved. Thus, the constraints set at a lower level of biological hierarchy confine higher levels of organization but may not explain or predict the more complex functions. Evolution is a "tinkerer" that combines the elements at hand together in new and unpredictable ways. The first "wave" in Figure 2.7 is close to the underlying chemical principles and will thus naturally represent a focus of this text.

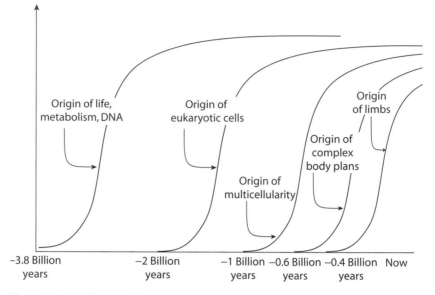

Figure 2.7: The history of the world according to cellular processes. Inspired by Marc Kirschner.

2.5   Summary

➤ Biological systems are defined by the interactions between their components.

➤ The links between molecular components are constrained by the basic laws of chemistry.

➤ Multiple links between components form a network, and the network can have functional states.

➤ Functional states of networks are constrained by various factors that are physicochemical, environmental, and biological in nature.

➤ The number of possible functional states of networks typically grow much faster than the number of components in the network.

➤ The number of candidate functional states of a biological network far exceed the number of biologically useful states to an organism.

➤ Cells must select useful functional states by elaborate regulatory mechanisms.

2.6   Further Reading

Danchin, A., THE DELPHIC BOAT: WHAT GENOMES TELL US, Harvard University Press, New Jersey (2003).

Harold, F.M., THE WAY OF THE CELL: MOLECULES, ORGANISMS AND THE ORDER OF LIFE, Oxford University Press, Oxford (2001).

Hartwell, L.H., Hopfield, J.J., Leibler, S., and Murray, A.W., "From molecular to modular cell biology," *Nature* **402**:C47–C52 (1999).

Gerhart, J., and Kirschner, M., CELLS, EMBRYOS, AND EVOLUTION: TOWARD A CELLULAR AND DEVELOPMENTAL UNDERSTANDING OF PHENOTYPIC VARIATION AND EVOLUTIONARY ADAPTABILITY, Blackwell Science, Malden, Mass. (1997).

Jacob, F., "Evolution and tinkering," *Science*, **196**:1161–1166 (1977).

Monod, J., CHANCE AND NECESSITY; AN ESSAY ON THE NATURAL PHILOSOPHY OF MODERN BIOLOGY, Knopf, New York (1971).

Nielsen, J., "It is all about metabolic fluxes," *Journal of Bacteriology*, **185**:7031–7035 (2003).

Papin, J.A., Reed, J.L., and Palsson, B.O., "Hierarchical thinking in network biology: The unbiased modularization of biochemical networks," *Trends Biochemical Sciences* **26**:641–647 (2004).

PART ONE

Reconstruction of Biochemical Networks

Cellular functions rely on the interactions of their chemical constituents. Various high-throughput experimental methods allow us now to determine the chemical composition of cells on a genome scale. These methods include whole genome sequencing and annotation (genomics), the measurement of the messenger RNA molecules that are synthesized under a given condition (transcriptomics), the protein abundance, interactions and functional states (proteomics) measurements of the presence and concentration of metabolites (metabolomic), and metabolic fluxes (fluxomics). In addition, methods now exist to determine the binding sites of protein to the DNA (location analysis) and to measure of a limited number of fluxes through reactions inside a cell. The physical location of protein products and segments of the DNA can be determined using various fluorescent reporting molecules. All these methods can be used to help to reconstruct the biochemical reaction networks that operate in cells. This part of the text discusses the reconstruction of metabolic, regulatory, and signaling networks. Given the rate at which new methods are being developed, it is likely that this part of the text will become dated the fastest. However, with new or existing methods the result of the reconstruction process is a set of chemical reactions or interactions that comprise these networks. The reader should be mindful of the fact that these are not separate and independent networks. In fact they interact with one another. We often tend to think of them as being separate based on the biases that the structure of the typical life science curriculum imposes. The recent discovery of the transcriptional regulatory roles of key glycolytic enzymes illustrates this point [113].

CHAPTER 3

Metabolic Networks

The function of cells is based on complex networks of interacting chemical reactions carefully organized in space and time. These biochemical reaction networks produce observable cellular functions. Network reconstruction is the process of identifying all the reactions that comprise a network. The reconstruction process for metabolic networks has been developed and implemented for a number of organisms. The main features of metabolic network reconstruction are described in this chapter. We briefly review the key properties of metabolic networks and introduce the hierarchical thinking that goes into the interpretation of complex network functions. Further details can be found in authoritative sources [120, 218].

As discussed at the end of this chapter, a true genome-scale reconstruction of cellular functions necessitates accounting for all cellular networks simultaneously. Such a comprehensive network reconstruction has yet to be established; therefore, in this chapter, we focus on metabolism and address the reconstruction of transcriptional regulatory and signaling networks in the following two chapters.

3.1 Basic Features

Intermediary metabolism can be viewed as a chemical "engine" that converts available raw materials into energy as well as the building blocks needed to produce biological structures, maintain cells, and carry out various cellular functions. This chemical engine is highly dynamic, obeys the laws of physics and chemistry, and is thus limited by various physicochemical constraints. It also has an elaborate regulatory structure that allows it to respond to a variety of external perturbations. Metabolic imbalance is involved in major human diseases, such as diabetes, obesity, cancer, and heart disease. Metabolism comprises two types of chemical

Table 3.1: Key chemical groups in metabolism and their carriers.

Group carried in activated form	Carrier molecule
Phosphoryl	ATP, GTP
Electrons	NADH, NADPH, FADH$_2$, FMNH$_2$
One carbon unit	Tetrahydrofolate
Methyl	S-Adenosylmethionine
Acyl (two carbons)	Coenzyme A, lipoamine
Aldehyde	Thiamine pyrophosphate
Carbon dioxide	Biotin
Nucleotides	Nucleoside triphosphates

transformations: *catabolic pathways* that break down various substrates into common metabolites and *anabolic pathways* that collectively synthesize amino acids, fatty acids, nucleic acids, and other needed building blocks. During these processes, an intricate exchange of various chemical groups and reductionoxidation (redox) potentials takes place through a set of carrier molecules (see Table 3.1). These carrier molecules and the properties that they transfer thus tie the metabolic network tightly together. Intermediary metabolism can be described at several levels of complexity (Figure 3.1).

Hierarchy in function of metabolic networks

Genome-scale reconstructions of metabolic networks contain hundreds of metabolites and sometimes over a thousand reactions (see Table 3.6). The functions of such networks are hard for the human mind to comprehend. We thus need mathematical models for the study of their properties and simulation of their function. However, as pointed out in Section 1.2, we can think of network properties in a hierarchical fashion to simplify the conceptualization of network functions. Such hierarchy can be based on manmade concepts, as discussed later, or can be the result of a nonbiased mathematical analysis of the stoichiometric matrix (see Chapter 9). In what follows, we briefly describe the traditional view of the hierarchical decomposition of the functions of metabolic networks (see Figure 3.1).

Level 1: Cellular inputs and outputs. Overall, intermediary metabolism comprises the enzymatic reactions pertaining to the transformation of substrate molecules into the essential building blocks of macromolecules and other vital products for growth and maintenance. A coarse-grained description of the overall activity of metabolism thus involves substrates as inputs and biomass and metabolic by-products as outputs. For industrial fermentation processes, a description of cells at this level has sufficed for

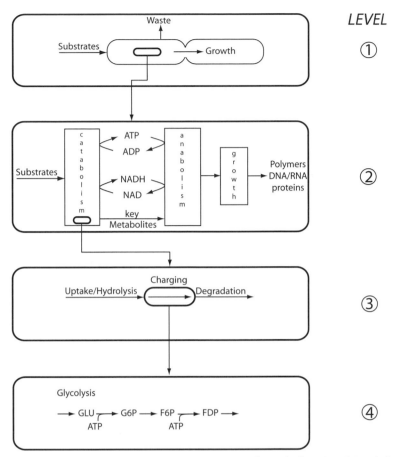

Figure 3.1: Four-level functional decomposition of metabolism. Level 1: whole cell; level 2: metabolic sectors; level 3: pathways; and level 4: individual reactions.

many purposes [6, 198]. The description comprises a simple set of coupled mass and energy balances, with various empirically determined "yield" coefficients that describe partitioning of the consumed substrate. Growth kinetics are given in terms of simple phenomenological models such as the Monod growth model. Models of this type are useful for a limited set of specific conditions. The yield coefficients are not constants; they change with the physiological state of the cell.

Level 2: Sectors. A bit finer grained look at intermediary metabolism reveals that it can be divided into two basic sectors (see Figure 3.2). Catabolism carries out the degradation of substrates via a series of converging pathways that lead to a set of 11 metabolites of central importance, called the *biosynthetic precursors*. Anabolism is a set of diverging pathways that originate from these central metabolites to form monomers or

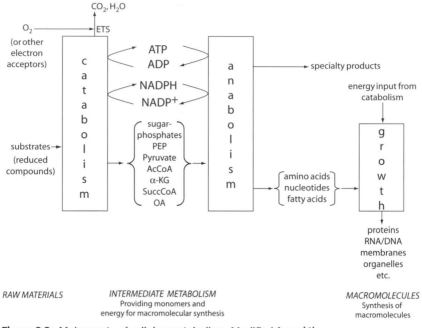

Figure 3.2: Major parts of cellular metabolism. Modified from [4].

building blocks for macromolecular biosynthesis. Genetically engineered bacteria used for bioprocessing, for instance, can be described at this level of complexity since it is appropriate for assessing host-plasmid interactions.

Level 3: Pathways. A still finer resolution reveals a situation in which pathways, and segments thereof, serve a definite role. For instance, catabolism of the major classes of biomolecules follows the same pattern; first, substrates are picked up by the cell, hydrolyzed if necessary, activated by a cofactor, and then degraded to yield energy and other properties stored on the carrier molecules. At this level of description, the essential features of metabolism begin to depend on basic chemical principles such as stoichiometric structure and kinetic regulation. Key metabolic pools, such as the *energy charge*, dominate the description, and key regulatory enzymes influence the motion of these pools and how mass and energy is distributed among them. There is currently much interest in the pathway level characterization of reconstructed biochemical reaction networks.

Level 4: Individual reactions. At the finest level of description one considers all the biochemical transformations that take place in a cell. Available high-throughput data, as discussed in Chapter 1, allows us to generate the information needed to describe cells at this resolution. It is at this level where this book is focused. We can now reconstruct genome-scale stoichiometric matrices of organisms and study them. The dimensions of these matrices are on the order of hundreds of metabolites and sometimes over a thousand chemical reactions, reflecting the complexity of a fully functional metabolic network.

Biochemical transformations fall into a few major categories. Some examples include transamination, phosphorylation, isomerization, dehydration, and dismutation. Thus, there are chemical "rules" that dictate what kind of links can exist in metabolic networks. As described later, biochemists have devised nomenclature that classifies these types of transformations and an Enzyme Commission (E.C.) number is associated with each enzymatically catalyzed metabolic reaction. Furthermore, there are thermodynamic restrictions associated with these transformations that dictate the energetic feasibility of a reaction and its equilibrium state. Thus, even though metabolic networks may appear complex, there are underlying physicochemical restrictions on their topological structure and network states. These constraints are detailed in Parts II and III of this book.

3.2 Reconstruction Methods

Defining the reaction list

The reconstruction of a genome-scale metabolic network relies on assembling various sources of information about all the biochemical reactions in the network. A variety of data sources can be used to synthesize a list of chemical reactions that form an organism's metabolic network (see Figure 3.3). The principal data sources are (roughly in the order of reliability) as follows:

1. *Biochemistry.* The strongest evidence for the presence of a metabolic reaction is found if an enzyme has been isolated directly from the organism and its function demonstrated. Extensive data is often available for model organisms, such as *Escherichia coli* and yeast but may be fragmented for organisms that have been sparingly studied.

2. *Genomics.* Functional assignments to open reading frames (ORFs), based on DNA sequence homology, may be used as a strong evidence for the presence of a reaction in an organism. Functional assignments can also be achieved from the genome location of an ORF and the cluster of genes that are found in its neighborhood. Genome annotations are subject to revision and updates.

3. *Physiology and indirect information.* Physiological evidence, such as the known ability of the cell to produce an amino acid **in vivo**, may

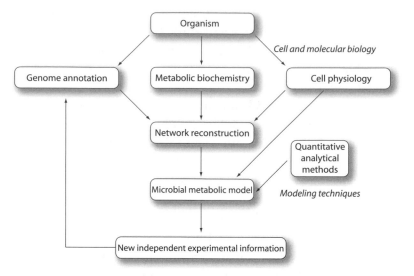

Figure 3.3: A schematic of the overall process of genome-scale metabolic network reconstruction (GENRE) and subsequent model formulation. Modified from [37].

lead us to include reactions which "fill in the pathway" to produce that amino acid. This process is called *gap analysis.* Other physiological information is often useful in diagnosing the function of a reconstructed networks.

4. *In silico modeling data.* Modeling and simulation studies often lead to the inclusion of metabolic reactions in the reconstruction. A network needs to be able to simulate cell behavior *in silico.* For instance, the metabolic network must be able to produce or take up all of the necessary components of the cellular biomass. One needs to add the reactions necessary to fulfill the biomass requirements if they are not present. Such reactions are referred to as "inferred reactions."

All the reactions identified by these various means then combine to produce a *genome-scale metabolic reconstruction* for the organism of interest. Normally, the reconstruction process starts with the annotated DNA sequence and thus the reconstruction is "genome-scale" since it will contain all the information that is found on the genome that relates to the organism's metabolism. This set of reactions comprises a *genome-scale metabolic model* when combined with quantitative analytical methods, which enable us to analyze, interpret, and predict integrated network functions (see Figure 3.3). Some of these mathematical methods that are scalable to the genome-scale networks are described later in the text.

Clearly, the confidence level in the various sources differs, but one can use quantitative scale to rank-order the reliability of the source. One such

Table 3.2: Methods for annotating genomes [108, 148].

ORF identification	"Traditional" annotation methods	New annotation methods
Stop codons	Experimental (direct)	Protein–protein interactions
GLIMMER	Sequence homology	Correlated mRNA expression levels
Genscan		Phylogenetic profile
		Protein fusion clustering
		Gene neighbors (operon clustering)
		Automation

quantitative scheme proposed is biochemical data (4), genetic data (3), genomic data (2), physiological data (1), and modelling data (0) [191]. One is never fully sure about the presence of a reaction until the biochemical data has been obtained, although sequence homology that meets certain criteria is often taken as sufficient evidence for a true functional assignment.

Genome annotation

Since few organisms have extensive biochemical information available, reconstruction relies heavily on an annotated genome sequence. ORFs are identified on the genomic sequence, then assigned a function. This process can be done through experimental methods (gene cloning and expression or gene knockout) or more commonly by comparing its sequence homology to genes of known function in other organisms. *In silico* annotation methods typically lead to functional assignment of 40–70% of identified ORFs on a freshly sequenced microbial genome. New and improved methods continue to be developed for genome annotation. For example, functions of gene products may be inferred from protein–protein interactions, transcriptomics, phylogenetic profiles, protein fusion, and operon clustering (see Table 3.2). It should be emphasized that every gene annotation based on *in silico* methods is hypothetical, and such annotation is subject to revision until the gene has been cloned, expressed, and the function of the gene product directly evaluated. The automation of network reconstruction from annotated sequence has been attempted [108]. To produce high-quality, well-curated reconstructions, one has to manually verify all the components and links in a network, since there are often subtle differences between even related organisms. There are many Web resources available for this purpose (Table 3.3).

Publicly available sources of sequence data

There are several publicly available databases that contain genomic data (Table 3.3). The Comprehensive Microbial Resource (CMR) provides tools

Table 3.3: Publicly available genome databases. Prepared by Ines Thiele.

Microbial genomes and annotation

DDBJ	http://www.ddbj.nig.ac.jp/
EBI	http://www.ebi.ac.uk/
EMBL	http://www.ebi.ac.uk/embl/
GenBank (NCBI)	http://www.ncbi.nlm.nih.gov/Genbank/
TIGR annotation software	http://www.tigr.org/software/

Comparative genomics

ERGO	http://ergo.integratedgenomics.com/ERGO/
The SEED	http://theseed.uchicago.edu/FIG/index.cgi
GenDB	http://www.cebitec.uni-bielefeld.de/groups/ brf/software/gendb_info/index.html
GeneQuiz	http://jura.ebi.ac.uk:8765/ext-genequiz/
MBGD	http://mbgd.genome.ad.jp/
Pedant	http://pedant.gsf.de/
Prolinks	http://128.97.39.94/cgi-bin/functionator/pronav
String	http://string.embl.de/
PUMA2	http://compbio.mcs.anl.gov/puma2/cgi-bin/index.cgi

Pathway/ Reconstruction tools

INSILICO discovery	http://www.insilico-biotechnology.com/f_products.html
MetaFluxNet	http://mbel.kaist.ac.kr/mfn
MFAML (Metabolic Flux Analysis Markup Language)	http://mbel.kaist.ac.kr/mfaml
SimPheny	http://www.genomatica.com/solutions_simpheny.shtml
Pathfinder	http://bibiserv.techfak.uni-bielefeld.de/pathfinder/
PATIKA	http://www.patika.org/

Pathway databases

BioSilico	http://biosilico.kaist.ac.kr or http://biosilico.org
KEGG	http://kegg.com/
MetaCyc	http://metacyc.org/
MRAD	http://capb.dbi.udel.edu/whisler/
Phylosopher	http://www.genedata.com/phylosopher.php
PUMA2	http://compbio.mcs.anl.gov/puma2/cgi-bin/index.cgi
EMP	http://www.empproject.com/

Enzymes

Brenda	http://www.brenda.uni-koeln.de/
KEGG	http://www.kegg.com/
IntEnz	http://www.ebi.ac.uk/intenz/

Proteins

HAMAP project	http://www.expasy.org/sprot/hamap/
InterPro	http://www.ebi.ac.uk/interpro/

Table 3.3 (continued)	
E. coli-specific	
Databases	
EcoCyc	http://ecocyc.org/
Colibri	http://genolist.pasteur.fr/Colibri/
GenProtEC	http://genprotec.mbl.edu/
CyberCell	http://redpoll.pharmacy.ualberta.ca/CCDB/index.html
EchoBase	http://www.ecoli-york.org/
Yeast-specific	
Databases	
CYGD	http://mips.gsf.de/genre/proj/yeast/
Saccharomyces Genome Database	http://www.yeastgenome.org/
H. pylori-specific	
Databases	
PyloriGene	http://genolist.pasteur.fr/PyloriGene/
hp-DPI	http://dpi.nhri.org.tw/protein/hp/ORF/index.php

for the analysis of 63 annotated genome sequences, both individually and collectively. The Institute for Genomic Research (TIGR) updates and maintains this site.

Another database that maintains many microbial genomes is the Genomes On-Line Database (GOLD) site. Not all of the information on the site is publicly available. The developers of GOLD have been active in automating the reconstruction of metabolic networks using pathway templates.

Biochemical data

Direct biochemical information is the most reliable source for the presence of a reaction in an organism. Biochemical data also gives stoichiometry and whether or not a reaction is reversible. For example, the enzyme that catalyzes the conversion of D-glucose to D-glucose-6-phosphate, as ATP is converted to ADP, is called glucokinase. The gene that encodes this enzyme is commonly called *glk*, and the E.C. number that corresponds to this reaction is 2.7.1.2. The structure of the Human β-cell glucokinase is shown at the top of Figure 3.5 (found in the Protein Data Bank). Collections of biochemical data on an organism's metabolism is often found in review articles and more recently in whole volumes that are focused on the biology of a single organism.

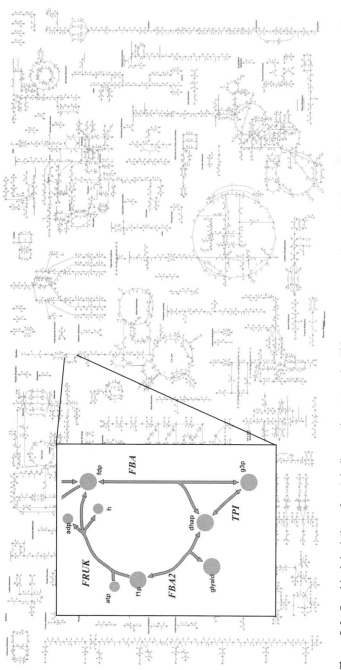

Figure 3.4: Graphical depiction of metabolic reaction networks. This map represents the metabolic network in yeast. Courtesy of Natalie Duarte; see http://systemsbiology.ucsd.edu for more details.

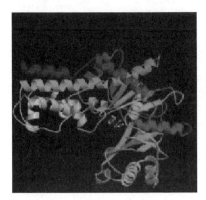

Gene: glk
Enzyme: Glucokinase
Reaction: ATP + D-Glucose =
           ADP + D-Glucose 6-phc
E.C.:   2.7.1.2

Figure 3.5: Biochemistry of the glucokinase reaction and and illustration of its protein structure. From http://www.rcsb.org/pdb/

Enzyme commission numbers

E.C. numbers are used to systematically characterize enzymatic reactions (http://www.chem.qmul.ac.uk/iubmb/enzyme/). They have been established to unambiguously classify reactions, which is needed because so many enzymes have ambiguous and duplicate names across organisms (see Table 3.4). For instance, try going to the E.C. Web site and searching first for succinate dehydrogenase (sdh), and then for fumarate reductase (fr). Both of these enzymes catalyze the same reaction, but in opposite directions. Some biochemists find that frd or sdh may be reversible at times. As a result, when you type in succinate dehydrogenase you will find that it is often used to indicate either reaction. A classification scheme similar to the E.C. system is being developed for transport reactions [26]. Unfortunately, there is no similar system for genes, which have the same problem of ambiguous and duplicate names. Thus, the curation of gene annotation information for a reconstruction can be quite laborious.

Protein databases

Swiss-Prot (http://us.expasy.org/sprot/) is a very useful source for examining particular protein or reaction assignments in detail and is considered a "gold standard" for biochemical information because it is so well-curated. It contains literature references, sequences, functional assignments, and other useful information, all specific to the organism being examined.

Table 3.4: Example of the E.C. nomenclature.

EC 1 Oxidoreductases	
EC 1.1	Acting on the CH—OH group of donors
EC 1.1.1	With NAD or NADP as acceptor
EC 1.1.2	With a cytochrome as acceptor
EC 1.1.3	With oxygen as acceptor
EC 1.1.4	With a disulfide as acceptor
EC 1.1.5	With a quinone or similar compound as acceptor
EC 1.1.99	With other acceptors
EC 1.2	Acting on the aldehyde or oxo group of donors
EC 1.2.1	With NAD or NADP as acceptor
EC 1.2.2	With a cytochrome as acceptor
EC 1.2.3	With oxygen as acceptor
EC 1.2.4	With a disulfide as acceptor
EC 1.2.7	With an iron–sulfur protein acceptor
EC 1.2.99	With other acceptors
EC 1.3	Acting on the CH—CH group of acceptors
EC 1.3.1	With NAD or NADP as acceptor
EC 1.3.2	With a cytochrome as acceptor
EC 1.3.3	With oxygen as acceptor
EC 1.3.5	With a quinone or similar compound as acceptor
EC 1.3.7	With an iron-sulfur protein acceptor

If one is not sure about the presence of a protein in an organism but a page is found for it on Swiss-Prot, he or she can be fairly sure that the protein has been characterized and that literature references are available. TrEMBL contains new entries to Swiss-Prot that have not yet been curated.

Gene–protein–reaction (GPR) associations

When associating genes to reactions, and vice versa, it is important to remember that not all genes have a one-to-one relationship with their corresponding enzymes or metabolic reactions. Many genes may encode subunits of a protein which catalyze one reaction. One example is the fumarate reductase. There are four subunits, *frdA*, *frdB*, *frdC*, and *frdD*, without which the enzyme (a protein complex) will not be able to catalyze the reaction. Conversely, there are genes that encode so-called *promiscuous enzymes* that can catalyze several different reactions, such as transketolase I in the pentose phosphate pathway. Such reactions typically involve similar chemical transformations of structurally related molecules.

These examples highlight the need to keep track of associations between genes, proteins, and reactions. Examples of different types of GPR associations are shown in Figure 3.6, where the top level is the gene locus, the

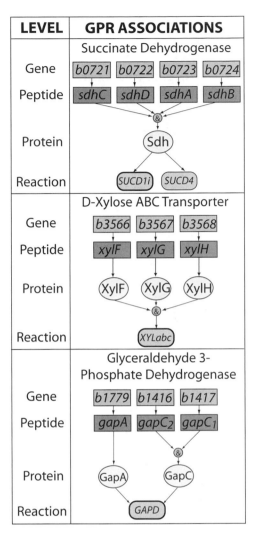

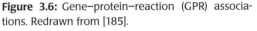

Figure 3.6: Gene–protein–reaction (GPR) associations. Redrawn from [185].

second level is the translated peptide, the third level is the functional protein, and the bottom level is the reaction. Many genes may encode subunits of a protein, or multiple proteins might come together to form an enzyme complex. Subunits (e.g., *sdhABCD* and $gapC_1C_2$) and enzyme complexes (e.g., *xylFGH*) are connected to reactions with "&" associations, meaning that all have to be expressed for the reaction to occur. For sdhABCD, the "&" is shown above the enzyme level indicating that all of these gene products are needed for the functional enzyme. With *xylFGH*, the "&" association is shown above the reaction level, indicating that the different proteins form a complex that carries out the reaction. Succinate dehydrogenase is an example of a promiscuous enzyme that can catalyze several different reactions. Isozymes (e.g., GapC and GapA) are independent proteins that carry out identical reactions. Only one of the isozymes needs to be present for the reaction to occur. Isozymes are shown as two or more arrows leaving different proteins but impinging on the same reaction.

Organism-specific sources of information

Several biological databases that integrate genomic and biochemical data for a particular organism are becoming available. One of the earliest of such sites is the *E. coli* encyclopedia (EcoCyc) database. Comprehensive Yeast Genome Database (CYGD), Yeast Protein Database (YPD), and *Saccharomyces* Genome Database (SGD) are some examples for yeast.

The widely used Kyoto Encyclopedia of Genes and Genomes (KEGG) database organizes its genomic information as maps of reaction networks. In reaction maps, arrows are used to connect various metabolites, indicating that one metabolite can be converted to another by a chemical reaction. This representation is the standard graphical representation of reaction and pathway data and will be described in Chapter 6.

For many organisms of interest, comprehensive textbooks have been written that include detailed descriptions of the organism's metabolism and other biological functions. These books give an overview of the organism's importance, metabolic features, and important references, as well as physiological data. The *E. coli* two-volume set [142] was the first of its kind and continues to be a useful source when building models of other bacteria. Several such organism-specific compendia have appeared [131, 138, 142, 211]. Such compilations of genetic, biochemical, and physiological data, and functional attributes of a particular organism, represent highly concentrated sources of data needed for reliable reconstructions.

In addition, to achieve a high-quality, well-curated network reconstruction, one should search the primary research literature. Comprehensive review articles are particularly useful since they contain organized collections of primary articles on a particular organism. Reviews are typically well summarized and written by experts on the subject and provide an accessible source of biochemical information. Frequently though, one has to search the primary literature, and searches may have to be done on a regular basis to continually update the network reconstruction.

Meeting demands and measured physiological states

There are two additional issues that one needs to consider in completing the reconstruction.

First, one needs to analyze the demands that are placed on the network. Most of the time, metabolic networks are fulfilling many functions, such as synthesizing the entire biomass composition of the organism in a growth state. In many cases, the biomass composition of an organism will not be available from direct experiments; for such cases, the biomass composition of a closely related organism may be used. For example, in reconstructing *H. pylori* or *H. influenzae*, one could assume a biomass composition similar to that for *E. coli*. This may not be an acceptable assumption for eukaryotes, such as *S. cerevisiae*. The best option is to experimentally determine the composition for the organism of interest. Knowing the relative macromolecular composition, such as the amino acid composition of proteins [233], is more important than detailed information on the makeup of each class of macromolecules.

Second, it is important to obtain physiological data to determine if the reconstructed network can reproduce physiological behaviors that have been observed experimentally. Such tests require integrated or mathematical descriptions of the network, detailed later in the text. Physiological data de facto gives the functional states of the network. The reconstruction must be able to reproduce these observed states.

Data on individual reactions and data on functional states represent fundamentally different information. The former is component-type information, often referred to as *bottom-up* data. The latter is whole network-type information, often referred to as *top-down* data. Since metabolic networks are functionally hierarchical, both these data types are important in obtaining genome-scale reconstructions.

Reconciliation and curation

Although a reconstructed network has been synthesized using various databases and literature sources for information, it is most likely not yet complete. Careful studies will often show that enzymes that are likely to exist in the thriving organism may be missing from the reconstruction. For example, both KEGG and TIGR give no indication that phosphofructokinase is found in *H. pylori*. This could mean that *H. pylori* is not able to produce 1,6-fructose bisphosphate (FDP) from glucose, although there may be other pathways by which FDP is produced. Careful review of the literature reveals that the phosphofructokinase enzyme may have been identified [90]. Other scientists, however, have disputed this claim. After thoroughly examining studies of *H. pylori* metabolism, one needs to decide whether or not to include this enzyme and the reaction it catalyzes in the reconstruction. Biochemical data is therefore fundamental to both curating and expanding the network.

Prospective design of experiments

No organism is fully characterized today. Therefore, although the online databases and all of the relevant literature have been searched and reactions tabulated, there is still a high probability that several necessary reactions will be missing in the reconstruction. Not all of the ORFs in the genome have been identified, assigned a function, and linked to reactions in the network. Based on knowledge of how an organism grows and functions, a gene product's presence can be inferred based on the inability of the organism to function without it. "Filling in the gaps" in this way is tantamount to stating hypotheses to drive further experimental research.

Indeed, the primary result from genome-scale constraint-based models of networks is a well-defined list of hypotheses and experiments to carry out in order to reconcile discrepancies in a reconstruction, fill in the gaps, and explore new functionalities of an organism. There are now a growing number of examples where models are used to drive such experiments, from well-characterized organisms, such as *E. coli* [51, 185, 33] and yeast [67, 47], to organisms that are not as well studied, such as *Geobacter sulfurreducens* [127]. This process of iterative model building promises to accelerate biological discovery, product development, and process design. It represents one of the major goals of systems biology [100, 152].

3.3 Genome-scale Metabolic Reconstructions

The reconstruction of metabolic reaction networks has been ongoing based on biochemical information de facto since the 1930s, when the glycolytic pathway was delineated. Since then, a large number of metabolic reactions have been discovered and described. Assembly of such reactions make up large sections of textbooks on biochemistry [120, 218]. Large-scale organism specific assemblies began to appear through multiauthored volumes in the late 1980s [142]. The availability of such information began the systematic synthesis of organism-specific metabolic networks. Large-scale reconstructions of *E. coli* metabolism were established in a stepwise fashion (see Table 3.5) and the network properties of their mathematical descriptions were assessed [183]. Similar biochemically based reconstructions have appeared for *S. cerevisiae* [54, 47] and *Aspergillus niger* [43].

The first genome to be fully sequenced was that of *H. influenzae* in 1995 [60], which enabled the first reconstruction of a genome-scale

Table 3.5: Pregenome era reconstructions of the metabolic network in *E. coli*. From [183].

Number of metabolites	Number of reactions	Publication
17	14	[129]
118	146	[231, 232]
305	317	[171, 172]

Table 3.6: Genome-scale reconstructions of metabolic networks in microbial cells. The detailed contents of many of these models are available at http://systemsbiology.ucsd.edu. Organelle-scale reconstructions of the human cardiac myocyte have appeared [238], accounting for 230 metabolites and 189 reactions. Compiled by Jennifer Reed.

Organism Organelle	Number of genes	Number of metabolites	Number of reactions	Publication
H. influenzae	296	343	488	[50]
E. coli	660	436	720	[51]
	904	625	931	[185]
H. pylori	291	340	388	[192]
	341	485	476	[221]
S. cerevisiae	708	584	842	[54]
	750	646	1149	[47]
G. sulfurreducens	588	541	523	[127]
S. aureus	619	571	640	[10]
M. succiniciproducens	335	352	373	[93]

metabolic network in 1999 [50]. Since then, a number of genome-scale reconstructions have been achieved (see Table 3.6). Most of the genome-scale networks reconstructed thus far are for bacteria, although the first genome-scale eukaryotic networks have recently appeared. Eukaryotic reconstructions are much more complicated than those of bacteria; for instance, the most recent *S. cerevisiae* reconstruction accounts for seven cellular compartments [47].

The process described in the last section represents the detailed lessons learned through these reconstruction efforts. The process of reconstruction is iterative. Unlike genome sequencing projects which have a well-defined end point, the reconstruction process is ongoing. The history of the reconstruction of the *E. coli* network is shown in Figure 3.7, and, at the publication of this book, reconstruction of this organism has been ongoing for close to 15 years. The reconstruction of other recently sequenced organisms is proceeding much faster now that a comprehensive reconstruction of the *E. coli* metabolic network is available.

3.4   Multiple Genome-scale Networks

Metabolic networks do not operate in isolation. They interact with many other cellular processes, such as transcriptional regulation and cellular motility. Signaling networks in multicellular organisms interact with metabolism, as do cellular fate processes, such as mitosis and apoptosis. To fully describe a cell, all these networks need to be reconstructed and

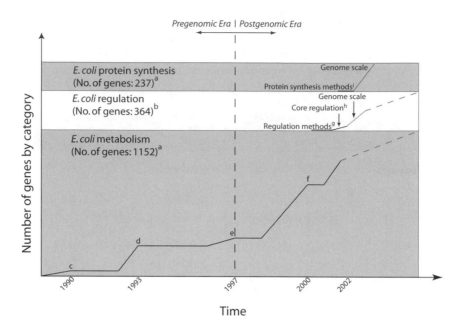

Figure 3.7: History of reconstruction of mathematically described metabolic reaction networks for *E. coli*. From [183].

integrated to simultaneously represent all cellular functions. Today, such integration has only been achieved for metabolism and transcriptional regulation [33]. As described earlier, genome-scale metabolic reconstructions are currently fairly comprehensive. Models of mitosis and apoptosis have appeared [223, 224, 68]. The status of reconstruction of transcriptional regulatory and signaling networks are described in the next two chapters.

Common components

The division of cellular networks into metabolism, regulation, and signaling has historical and life science curriculum origins. However, there are an increasing number of discoveries showing that often the same molecules participate in more than one of these networks (see Table 3.7). We must therefore begin to think of all the chemical reactions resulting from the activities of genomes and gene products as one genome-scale network.

Putting "content in context"

Over the coming decade we may expect to see reconstructions appear that integrate multiple such networks. As in the past, it is likely that model organisms such as *E. coli* will lead the way (see Figure 3.8). Seen from a broader perspective, reconstructing genome-scale models provides a

Table 3.7: Cellular components of multiple networks (signaling, metabolism, and regulation). From [157].

Component	Metabolic	Regulatory	Signaling	Ref.
ATP	Energy metabolism	Global regulator of DNA coiling	Phosphate is ubiquitous in signaling reactions	[83]
Riboswitches and ribozymes	Metabolite-binding can regulate activity	Ribozymes cleave RNA transcripts; control gene expression		[250]
Arg5,6	Arginine biosynthesis	Binds to specific nuclear and mitochondrial loci of DNA		[81]
Phosphoinositides	Lipid biosynthesis		PI3K signaling	[27]
GAPDH	Glycolytic enzyme	Transcriptional coactivator		[208]
Lactate dehydrogenase	Glycolytic enzyme	Transcriptional coactivator		[208]
Nicotinamide adenine dinucleotide (NAD)	Metabolic cofactor	Alter TF DNA-binding properties	Calcium signaling; Poly-ADP-ribose polymerases	[208, 14]
Interleukin-1	Inhibition of fatty acid synthesis		Immune system signaling	[133]
Sialic acid	Oligosaccharide synthesis		Apoptotic signaling	[140]
Insulin	Glucose uptake		PI3K signaling	[133, 206]
Hog1	Glycerol synthesis; phosphofructokinase		Osmolarity response in yeast	[45]

formalism for integrating all of the "omics" data that is currently available, or allows one to put "content in context." Here, genomic, transcriptomic, proteomic, and metabolomic data are all integrated in the context of a biochemically and genetically accurate framework that enables one to make predictions about whole organism function, given the nutritional environment.

Note that chemical composition data (genomics, transcriptomics, proteomics, metabolomics) and component interaction data (DNA-protein and protein–protein interactions, or "interactomics") can be comprehensively

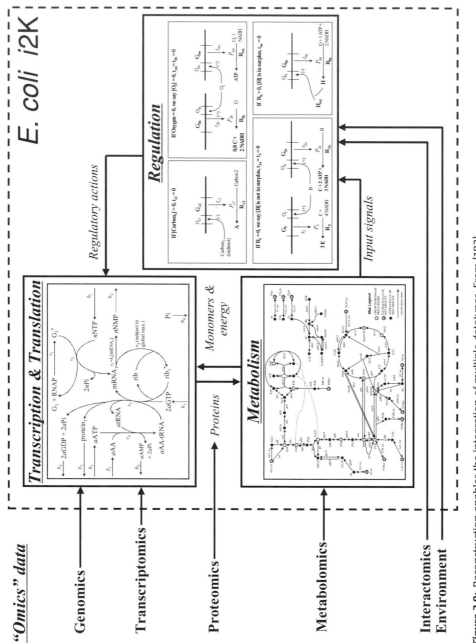

Figure 3.8: Reconstruction enables the integration of multiple datatypes. From [183].

Table 3.8: The data types accounted for in genome-scale multinetwork reconstruction.

Genomic data	Annotated genes
	Splice variants
	Gene location
	Regulatory regions
	Wobble base pairs
Biochemical data	Stereochemistry (L and D forms)
	pH and pK_a to determine charge
	Elemental balance and charge balance
	Multiple reactions/enzyme
	Multiple enzymes/reaction
Transcription/translation	Gene–transcript–protein–reaction association
	Transcript and protein half-lives
	tRNA abundances
	Ribosomal and polymerase capacities
Physiological data	Fluxes (fluxomic data)
	Overall phenotypic behavior
	Gene knockout phenotypes
	Compartmentalization of gene products

connected to phenotypic data (network functional states, such as flux-omics and growth rate phenotyping). Because of the predictive nature of mathematical models, they can also be used to to curate and critically examine high-throughput data by reconciling *in silico* predictions with experimental results. Thus, the reconstruction can be called a *biochemically, genetically, and genomically structured (BIGG) database*, and the mathematical analysis approach can be called *query* or *interrogation* methods.

Data types accounted for in a multinetwork reconstruction

The reconstruction and integration of multiple networks will allow for the simultaneous accounting of diverse date types and their reconciliation. Genomic, biochemical, macromolecular, and physiological data can be used in such reconstructions (Table 3.8).

- Genomic data include DNA sequences and the location and functional annotation of genes. Transcriptional regulatory models also account for many of the intragenic regions where RNA polymerase and transcription factors bind.
- Extensive biochemical information is contained in reconstructions. The L and D forms of compounds are accounted for separately. A molecule's charge can be determined from its pK_a value and network-scale proton balancing. All reactions must elementally and

charge balanced. Promiscuity of enzymes and the ability of different enzymes to catalyze the same reaction needs to be accounted for.

- Reconstructions of translation and transcription include the relationship between an open reading frame and its transcript. Translation not only associates a transcript with a protein but also enables the incorporation of transcript half-life data, tRNA abundances, and ribosomal capacities. Wobble base pairs can also be associated with corresponding tRNAs. The assembly of multiple proteins to form functional complexes can also be incorporated.
- Large-scale reconstructions allow us to simulate and thus reconcile phenotypic data. Fluxomic data give information about the actual flux distributions in a network and can be derived from a mathematical model. The consequences of removing a gene can be assessed. The cellular location of proteins can be described, as in the seven-compartmental model of yeast.

Genome-scale reconstructions provide a mechanistic framework for the integration of a wide range of data types. Such reconstructions, and their stoichiometric representation, are a common denominator in systems biology (recall Figure 1.4).

Regulation of metabolic networks

Regulation of metabolism is accomplished by modulating enzymatic reaction rates. Such modulation is achieved by either regulating the activity or the concentration of key enzymes. In some cases, both kinetic function and enzyme concentration are regulated. These two types of regulation are illustrated in Figure 3.9, where feedback loops at the end of a pathway regulate the first reaction in the sequence and also control the production rate of the enzymes catalyzing reactions in the pathway. Regulation on both levels can be either

- *negative*, called *repression* in the case of regulation of gene expression and *inhibition* in the case of regulation of enzyme activity, or
- *positive*, called *induction* in the case of gene expression and *activation* in the case of regulation of enzyme activity.

The time scale of regulation of enzyme activity is typically much shorter than that of gene expression, that is, on the order of minutes and hours, respectively [75]. Normally, regulation of gene expression is considered a coarse control of metabolism, whereas regulation of enzyme activity is viewed as fine tuning. Regulation of gene expression is fairly well-characterized in bacteria, but much more complicated regulatory patterns are found in eukaryotic cells. A detailed account of the principles of metabolic regulation is given in [84]. Transcriptional regulation is detailed

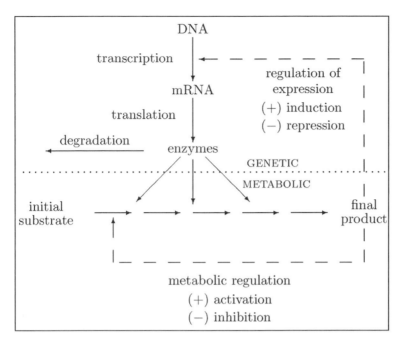

Figure 3.9: schematic of interacting metabolic and genetic control loops, modified from [151]. The dashed lines indicate regulatory interactions, while the solid arrows indicate primary chemical transformations. The dotted line is used to seperate the domain of metabolic from that of transcriptional regulation.

in the next chapter, but the regulation of gene product activity is briefly discussed here.

Regulation of enzyme activity

Cells use various mechanisms to regulate the activity of enzymes. For instance, many key enzymes in metabolism are regulated by an *allosteric mechanism*. In addition to having the binding site for the substrate, allosteric enzymes have a binding site for regulatory molecules as well. A bound regulatory molecule can either activate or inhibit enzyme activity. Allosteric interactions between the catalytic and regulatory sites cause conformational changes in the enzyme molecule. This is indicated by the name allosteric, coined by [139], which means different (allo-) binding sites, as opposed to isosteric, where substrates and modulators would bind to the same site. Allosterism is quite advantageous since the substrate and regulatory molecules are typically chemically unrelated.

Allosteric enzymes are commonly found at the beginning of a sequence, whereas their regulators are found at the end. This forms feedback loops, as indicated in Figure 3.9. Hexokinase, the enzyme catalyzing the phosphorylation of glucose, is a familiar example. It is inhibited by the chemical energy source ATP, which is one of the primary products of glycolysis, and is stimulated by ADP, which is produced by consumption of the energy stored in ATP and may be considered a substrate. Allosteric regulation and other regulatory mechanisms, such as protein phosphorylation, have an underlying chemical mechanism and can thus be described stoichiometrically.

3.5 Summary

➤ Complex networks operating inside cells carry out complicated biological functions. Examples include metabolic, regulatory, and signaling networks.

➤ All networks are based on underlying biochemical reactions and can thus be described by a stoichiometric matrix.

➤ Reaction networks can be described at different levels of resolution enabling us to conceptualize their functionalities in a hierarchical fashion.

➤ Metabolism is the best characterized cellular reaction network in terms of its biochemistry, kinetics, and thermodynamics.

➤ Genome-scale reconstruction of metabolic networks for organisms whose genome has been sequenced is now possible.

➤ Network reconstruction is a detailed, laborious process that needs careful examination of all the components and links in the network. Procedures to perform this task have been developed. Numerous Web resources and tools are available to aid in developing curated networks.

➤ Metabolic networks interact with essentially all other cellular processes. The reconstruction of these processes and the integration of multiple networks will lead to the description of a comprehensive range of cellular functions.

➤ Such a multinetwork reconstruction represents a biochemically, genetically and genomically, structured database that provides the framework for analyzing -omics data types.

3.6 Further Reading

Covert, M.W., Schilling, C.H., Famili, I., Edwards, J.S., Goryanin, I.I., Selkov, E., and Palsson, B.O., "Metabolic modeling of microbial stains *in silico*," *Trends in Biochemical Sciences*, **26**:179–186 (2001).

Lehninger, A.L., PRINCIPLES OF BIOCHEMISTRY, Worth Publishers, New York (1993).

Palsson, B.O., "*In silico* Biology through 'Omics'," *Nature Biotechnology*, **20**:649 (2002).

Reed, J.L., and Palsson, B.O., "Thirteen years of building constraints-based *in silico* models of *Escherichia coli*," *Journal of Bacteriology*, **185**:2692–2699 (2003).

Stryer, L., BIOCHEMISTRY, W.H. Freeman, New York (1997).

CHAPTER 4

Transcriptional Regulatory Networks

The expression of the gene complement of a genome is carefully regulated. Only a fraction of the genes in a genome are expressed under a given condition or in a particular cell type. There is a complex transcriptional regulatory network that controls which genes are expressed in response to various environmental and developmental signals. Extensive effort is being devoted to the elucidation of the components of transcriptional regulatory networks and the links between them. The reconstruction methods that are being developed are based on both legacy and high-throughput data types, and notable progress is being made with a few specific cases. Although the comprehensive details are not yet available for any one transcriptional regulatory network, some of their fundamental principles have been elucidated and a conceptual framework for their hierarchical decomposition has been developed.

4.1   Basic Properties

The chemical conversions taking place in metabolic networks relate to the dismemberment and assembly of small molecules through a series of chemical transformations. In contrast, transcriptional regulatory networks involve the association and interaction of large molecules. They rely primarily on protein–protein interactions and DNA–protein interactions, although metabolites do participate directly in some of these transformations. The chemistry underlying these interactions is currently partially understood, but much progress is being made. Still, however, these networks are not as well assembled and characterized as metabolic networks.

Some of the key features of regulatory networks are emerging. Specificity in regulatory networks is achieved by specificity in binding and association of macromolecules, often in particular locations in the cell.

Once the components are colocalized, specific interactions among them can take place. The localization step is critical, and it is governed by collision frequency, or mass action kinetics, of the participating components. The strength of the association is determined by the chemical composition and molecular structure of the surfaces of the interacting macromolecules. The extent of flexibility in determining the properties of these surfaces by changing the amino acid sequence in proteins and the base sequence in DNA binding sites is unknown [21]. This process has been termed *regulated recruitment* [180] and the binding affinity, or the "stickiness," of the complementary surfaces is of key importance. Once we better understand the constraints and limitations of this process, we will understand how easily new links can form or old links can disappear in transcriptional regulatory networks.

To get the reader oriented, two examples that are of historical importance are provided. These two examples illustrate the effect of the DNA-binding proteins alone and in combinations on the transcription of the target gene. The activity of the DNA-binding proteins themselves is controlled by various signaling pathways, which are not described in these examples.

The *lac* operon in *Escherichia coli*

The *lac* operon consists of three structural genes (*lacA, lacZ, and lacY*) involved in lactose utilization. The operon is regulated by lactose and glucose signals mediated by two DNA-binding regulatory proteins: the *lac* repressor and CAP, respectively. The *lac* repressor binds DNA only in the absence of lactose, whereas CAP binds only in the absence of glucose. Depending on the lactose and glucose concentrations in the medium, three different states for the regulatory system can be identified (Figure 4.1):

1. If both lactose and glucose are present, neither *lac* repressor nor CAP is bound to DNA, RNA polymerase binds weakly to the promoter, and the operon is transcribed at a low basal level.

2. If lactose is present and glucose is absent, CAP is bound to the promoter, but *lac* repressor is not bound to its site. CAP will now preferentially recruit RNA polymerase to the promoter, and the expression level of the lac genes is increased 40-fold.

3. If lactose is absent (independent of whether glucose is present or not), *lac* repressor is bound to DNA. Because the binding site of the *lac* repressor is within the operator, it excludes the RNA polymerase from the promoter, and the expression of the lac genes is strongly repressed.

A fourth state of this system at low lactose in the absence of glucose has been described recently [205].

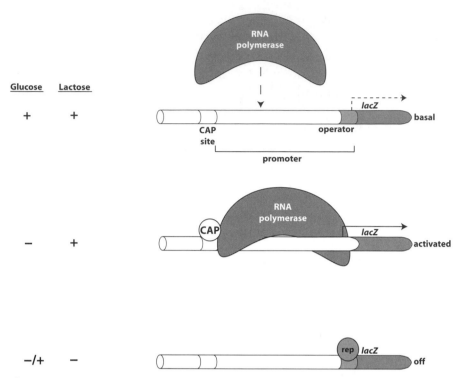

Figure 4.1: Transcriptional regulation of the *lac* operon in *E. coli* by the *lac* repressor (rep) and CAP in response to the presence of glucose and lactose. The logistical representation can be formulated based on these experimental observations. Redrawn from [180].

The *GAL* regulon in yeast

The *GAL* genes are required for the breakdown of galactose, a sugar hexose. Their transcription is induced by galactose and repressed by glucose. Galactose induction is mediated by a DNA-binding activator, Gal4, and repression by a DNA-binding repressor, Mig1. Similar to the *lac* operon, three different binding, and thus functional, states of this regulatory system can be identified (Figure 4.2):

1. If galactose and glucose are both absent, Gal4 is bound to its binding site as a homodimer, and Mig1 is not bound. However, Gal4 is not able to recruit the polymerase and activate transcription because it is complexed with the Gal80 protein, which acts as an inhibitor.

2. If galactose is present and glucose is absent, Gal80 inhibition is released and Gal4 recruits the RNA polymerase to the promoter. Due to the low basal level of transcription in eukaryotic cells, Gal4 induces GAL gene transcription more than 1000-fold.

3. If galactose and glucose are both present, both Gal4 and Mig1 are bound to the promoter. However, the activating effect of Gal4 is

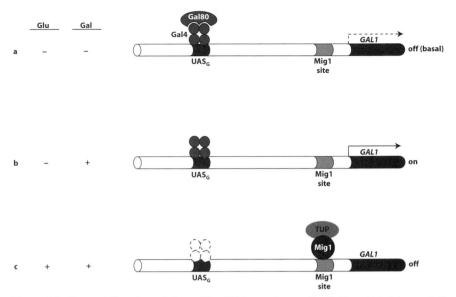

Figure 4.2: Transcriptional regulation of the *GAL1* gene in yeast by Gal4 and Mig1 transcription factors in response to glucose (Glu) and galactose (Gal) signals. Redrawn from [180].

counteracted by the much stronger repressing effect of Mig1, resulting in strong repression of *GAL* gene transcription. Mig1 acts by recruiting a corepressor complex, which represses transcription by an ill-characterized mechanism.

Note that the underlying experimental facts in these two examples are described essentially in terms of logistical statements. Such statements can be mathematically represented. Of course in reality, these different binding states represent chemical events that are determined by concentrations and binding affinities. If the chemical equations describing DNAbinding of regulatory proteins are known, these regulatory circuits can be described stoichiometrically. Furthermore, given the small number of some of the regulatory molecules and the thermal noise that exists inside cells, stochastic factors are believed to play an important role in the kinetic behavior of regulatory networks, at least at the lower end of their hierarchical scale [107, 182].

Proteins that bind to DNA

E. coli DNA-binding proteins are classified into several major categories: DNA packaging, DNA recombination, DNA repair, DNA replication, transcription initiation, RNA synthesis, and transcriptional regulation (see Table 4.1). Most of the proteins involved in transcription regulation and

Table 4.1: Functions and examples of *E. coli* DNA-binding proteins.

DNA packaging	Nucleoid proteins (OmpH)
DNA recombination	DNA strand exchange, renaturation proteins (RecA)
DNA repair	uracil-DNA glycosylases (Ung), DNA endonucleases (Vsr)
DNA replication	Origin binding proteins (Rob), DNA polymerases (Po1A), DNA ligases (LigA), single-strand binding proteins (Ssb), DNA topoisomerases (GyrA)
Transcription initiation	Subunit of RNA polymerase (RpoD)
RNA synthesis	RNA polymerases (RpoA)
Transcription regulation	186 transcription factors, including Crp, Fnr, ArcA, Fis, HimA, PhoB, etc.
Others	Restriction enzymes (McrA)

DNA repair recognize specific DNA sequences and bind predominantly to these target sequences, whereas others may bind nonspecifically at various positions along an *E. coli* genome.

The copy number of RNA polymerases can be related to a degree of expression activity in a cell. In a fast-growing *E. coli* cell, there are roughly 3,000 RNA polymerase complexes. Most transcription factors recognize upstream promoter sequences and upregulate or downregulate transcription initiation at the promoter to which these elements are attached. According to RegulonDB (http://www.cifn.unam.mx/Computational_Genomics/regulondb/), there are 186 known and 141 additional predicted transcription factors in *E. coli* [165].

Each gene (or operon in prokaryotes), in a genome has at least one promoter region associated with it, similar to the ones described in the earlier examples. Genes whose expression is controlled by a set of DNA-binding proteins binding to their promoter region can in turn act as regulatory proteins. When all of the DNA-binding proteins and their target promoters in a genome are considered together, a complex regulatory network with transcriptional cascades and feedback loops emerges. The earlier examples illustrate fairly simple cases of regulation with only two to three different DNA-binding proteins acting on one promoter. In higher eukaryotes, the complexity of promoter and enhancer regions of the genome is usually much higher, and these regions can contain binding sites for tens of different regulatory proteins. The higher the number of molecules participating in transcriptional regulation, the larger the combinatorial possibilities are, and thus a larger number of functional states can be derived as the number of components grows.

(A) Regulatory networks can be decomposed into a small set of commonly occurring structural "motifs"

(B) Behavior of prototypical examples of these motifs can be studied to gain insight into their role in the full regulatory network

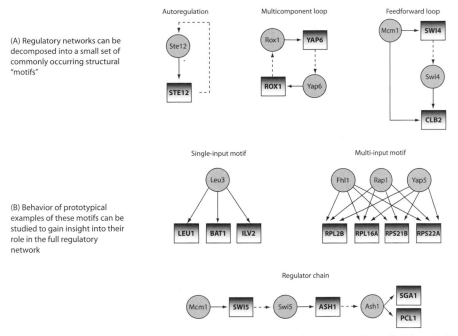

Figure 4.3: Motifs in transcriptional regulatory networks: (A) adapted from [207] and (B) adapted from [188]. Redrawn from [207].

Fundamental building blocks

Although the fundamental reaction chemistry is the same for metabolic and regulatory networks, the types of reactions that form the "building blocks" of transcriptional regulatory networks are different than those of metabolism. The basic functional block of a regulatory network is the promoter region of a gene or operon, which contains the *cis*-regulatory binding sites for the relevant transcription factors regulating the expression of a particular gene, as illustrated by the earlier two examples. The locations and orientations of these binding sites, as well as the affinity of the transcription factors to particular variants of the site, determine the expression levels of a gene in response to changes in the active transcription factor concentrations.

The transcriptional regulatory network is then defined by which transcription factors bind to which promoters and what the integrated effect of all these transcription factors is on the expression of genes [180]. It has been demonstrated that the known organization of promoter regions in bacteria allows for the implementation of a wide class of regulatory logic functions within a single promoter [141], such that a single "node" in the regulatory network can be relatively complex. Regulatory networks can be decomposed into a small set of commonly occurring structural "motifs" [207], summarized in Figure 4.3. The behavior of prototypical examples of these

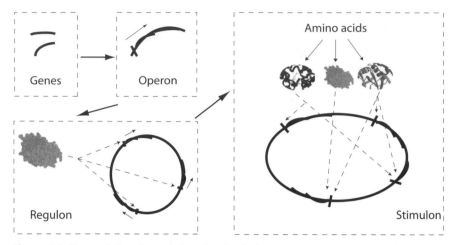

Figure 4.4: Transcriptional regulation: levels of abstraction by gene, operon, regulon, and stimulon.

motifs can be studied to gain insight, in a bottom-up fashion, into their role in the full regulatory network [188, 251].

Hierarchy in transcriptional regulatory networks

As for metabolism, there are several levels of abstraction to consider for transcriptional regulatory networks (see Figure 4.4). The simplest elements are the genes. Some genes are *constitutively* expressed, meaning that they are always transcribed at relatively constant levels. Other genes are regulated, and the transcriptional regulatory network induces or represses transcription of these genes. A group of genes that are adjacent on the genome and are transcribed together forms a unit called an *operon*. This arrangement is primarily found on prokaryotic genomes. An additional means of regulating related genes as a group is a *regulon*, where a certain regulatory protein binds to multiple locations on the DNA, causing induction or repression (or a combination of both) of the related genes or even multiple operons. This mode of coordination is the method of choice for eukaryotes.

At the highest level of abstraction is the *stimulon*, which includes all the regulons that are induced by a particular stimulus. For example, a stimulon may be all the regulons that are related to a particular substrate. If that substrate, such as an amino acid or glycerol, is present in the extracellular medium, representing a stimulus sensed by the cell, it will result in the activation or inactivation of certain proteins in the cell. These proteins in turn have an effect on the transcription of various genes and operons, resulting in a new set of available metabolic or other types of proteins in the cell. The end result is an altered behavior. In the case of amino acids, they are most likely no longer synthesized by the cell de novo, but are instead taken up from the environment. In the case of glycerol, there are five operons on the *E. coli* genome involved in its metabolism. They all react in a coordinated manner to the presence of glycerol in the medium.

The spatial or topological configuration of the genome influences gene expression. Such effects can be global or local. For instance, the *energy charge* itself is a global regulator of gene expression through its alteration of genome geometric arrangement [83]. Thus, in addition to the network of specific binding events that place the RNA polymerase throughput the genome, there are three-dimensional aspects to the expression state of the genome. The reader should recall Figure 2.3 that conveys the intricate organization and crowding of the cell's interior, within which genomes function.

4.2 Reconstructing Regulatory Networks

The magnitude of the task

Estimating the scope of a metabolic network reconstruction task for a given organism can be done relatively easily by estimating the number of genes with potential metabolic function present in the genome. This number is based on its annotation. For regulatory networks, the number of transcription factors can not simply be used to estimate the complexity of the network since the transcription factors can have multiple target genes and often act in synergistic combinations. However, the relative fraction of transcription factor coding genes tends to be higher for organisms that encounter more varied environmental conditions during their lifetime [29], indicating that there are limits to the range of transcriptional states that can be achieved with a fixed number of transcription factors.

Information on well-studied organisms can be used to evaluate the level of complexity of transcriptional regulatory networks in terms of the number of components (TFs, target genes) and regulatory interactions (Table 4.2). *E. coli* has been predicted to have 314 transcription factors [165], and based on primary literature, 577 regulatory interactions have been identified [207]. In yeast *Saccharomyces cerevisiae*, there are 203 verified or putative DNA-binding transcription factors, and large-scale protein-DNA binding screens indicate that there are at least 3,500 high-confidence regulatory interactions [82]. For both *E. coli* and yeast these numbers of regulatory interactions are most likely underestimates [33], but they give an indication of the order of magnitude of the regulatory network reconstruction task. Although the numbers of regulatory interactions appear to be large, developments in experimental techniques as well

Table 4.2: Reconstructed regulatory network structures in *E. coli* and *S. cerevisiae*. Prepared by Markus Herrgard.

Network	Regulatory genes	Target genes	Regulatory interactions	Regulated reactions
E. coli full metabolic [33]	104	451	–	555
E. coli database [207]	123	762[a]	1468[a]	–
S. cerevisiae metabolic [88]	55	348	775	–
S. cerevisiae database [80]	109	418	945	–
S. cerevisiae ChIP-chip [82]	203	1296[b]	3353[b]	–

[a]Counting each gene in an operon separately.
[b]Includes only high confidence interactions.

as computational methods make genome-scale regulatory network reconstruction a feasible task, at least for well-studied microbial organisms.

Three fundamental data types

As conceptually described in Chapter 2, systems biology is about components, how they are linked together to form networks, and the functional states that these networks take. Consequently, there are three data types of interest:

1. *Component data.* We can break cells apart, then isolate and identify their components. For transcriptional regulatory networks, the data types in this category include the identification of binding sites, the transcription factor molecules, riboswitches, and so forth. Significant relevant legacy data exist, and so do ORF functional assignment data. Both are needed to determine the scope the reconstruction effort.

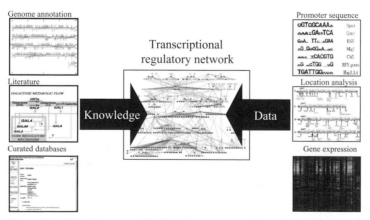

Figure 4.5: Data types used in the bottom-up and top-down reconstruction approaches to transcriptional regulatory networks. Courtesy of Markus Herrgard.

2. *Interaction data.* Links are formed by chemical interactions between components. There are many methods, both experimental and computational, being developed to determine such interactions, including DNA–protein, protein–protein, and metabolite–RNA interactions. Global data sets often have significant error rates, and ideally all such interactions should be directly verified by small-scale experiments. Furthermore, a set of positive and negative controls should always be included.

3. *Network state data.* The reconstructed networks have functional states. The state of a whole network can be assessed by a variety of data generated from living cells in well-defined environments, such as genome-scale expression data and phenotyping data. Controls for network states are often assessed through *perturbation experiments* (see Chapter 12). Network perturbation experiments include genetic perturbations (gene knockouts, gene silencing through inhibitory RNA), environmental perturbations (changing the availability of nutrients, shocking cells by changing temperature or pH), systemic perturbation (through adaptation), and diseased states (normal vs. pathological).

The integration of component, link, and network state data is needed for network reconstruction and validation.

Top-down data types
High-throughput data types that simultaneously measure a large number of variables or states are often referred to as *top-down* data. We discuss three such top-down methods:

1. *Experimentally determining the expression state of a genome.* Genome-scale mRNA expression profiling is perhaps the most common of such data types. Such data give the expression level of potentially every gene being expressed in an organism under a particular condition. One can then use gene knockouts to remove a transcription factor from the genome and then expression profile the knockout organism under the same condition. Then a comparison of the two expression states allows one to infer the role of the missing transcription factor in the regulation of genes whose expression is altered. Such experiments are known as *genetic perturbation* experiments. This approach has been used to study the GAL operon in yeast [100] and the oxygen shift in *E. coli* [33]. Such inference of regulatory interactions can also be achieved using other approaches [97, 202, 241].

2. *Identifying all promoter sites using computational approaches. Promoters* are genomic locations near the transcriptional start sites

of genes. The knowledge that transcriptional regulatory proteins normally act by binding semi-specifically in promoter regions has spurred the development of numerous *in silico* methods to identify promoter sites.

One class of analyses is aimed at discovering the actual DNA sites where regulatory proteins bind. Some algorithms work by considering all known promoters in an organism and searching for similar nucleotide sequences that occur significantly more often than would be expected by chance. Since different regulatory proteins often act synergistically, other methods exploit this fact by searching for different groups of similar nucleotide sequences that occur near each other an improbably high number of times.

Expression data can be combined with such *in silico* search methods. Genes are first clustered based on similar expression patterns, and then one can infer that a similarity of expression implies similarity of transcriptional regulation. This similarity in regulation often means that one or more regulatory motifs will be enriched in the clustered promoters. But due to the combinatorial nature of interacting regulatory proteins, this assumption may not always be completely true.

3. *Experimentally determining the location of protein binding sites on DNA.* The inference of a regulatory interaction based on *in silico* methods needs to be examined by a direct experiment. High-throughput data sets can lead to the generation of numerous such inferences. Thus, high-throughput methods are needed to test, validate, and refute such inferred interactions. One such approach is to use the so-called genome-wide location or ChIP–chip analysis [187](see Figure 4.6). In this approach, the transcription factors are cross-linked to DNA under the physiological condition of interest. The DNA is isolated and fragmented by sonication. Then an antibody specific to a particular transcription factor is used to isolate the transcription factor and the DNA fragment to which it is bound. Following such isolation, the DNA fragment is released from the protein and its sequence is identified by hybridization to a microarray containing promoter sequences, and the relative amount of binding can be quantified. The sequence of the DNA fragment can then be compared to the genomic sequence to identify the binding site(s) of the transcription factor.

An advantage that regulatory network reconstruction has over metabolic network reconstruction is the availability of high-throughput experimental data directly relevant to the network structure. For metabolic processes, the

A method for identifying target genes for TFs at the genome scale

Combines chromatin immuno-precipitation (ChIP) with intergenic DNA microarrays for signal detection

Each TF–promoter pair gives a probablility that the TF binds the promoter under a particular condition

Has been applied to 106 TFs in yeast under one condition

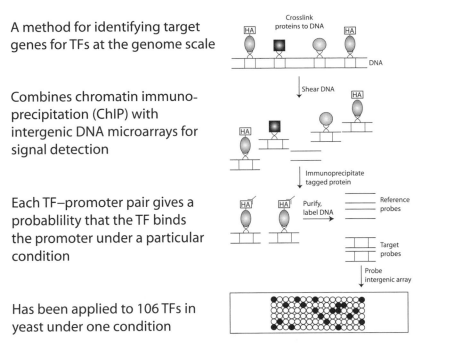

Figure 4.6: Location or ChiP–chip analysis. Modified from [116].

only widely available data source is the genome sequence. Unfortunately, methods for measuring relevant metabolic quantities such as metabolic fluxes and metabolite levels are still not commonly used at the genome scale [216]. On the other hand, the two primary data types useful for the regulatory network reconstruction task – genome-wide mRNA expression and location analysis data – are widely available.

Bottom-up data types

Data derived from classical biochemistry or genetics that are focused on a single or a few variables are often referred to as *bottom-up* data. Such data on individual transcription factors can be found in the literature. In some cases, such as for the transcription factors FNR or ArcA in *E. coli*, such literature searches would yield a large number of primary research reports, and often review papers. In other cases, little information is found in the primary literature.

As the *lac* operon and GAL regulon examples show, there are select cases in which the regulatory structures of operons or regulons have been eluci-dated. Such data may be available for select model organisms. Accumula-tion of such data has resulted in the construction of databases that target transcriptional regulation. One prominent example is the RegulonDB that contains information about 186 transcription factors in *E. coli*. There are also general databases for individual organisms such as YPD for yeast [40] that contain significant amounts of regulatory information. In addition to databases describing regulatory network structures, there are also comprehensive databases specializing in describing transcription factor binding sites such as SCPD [261] for yeast and the general transcription factor binding site database TRANSFAC [134]. Although these databases contain valuable information for regulatory network reconstruction, they are not very complete and for the most part lack information about the synergistic effects between transcription factors acting on one gene. Nevertheless, these databases and primary research literature can be utilized to reconstruct regulatory networks for well-characterized organisms such as *E. coli* and yeast [35, 207].

The bottom-up approach to reconstruction is laborious, as one has to study the network component by component. Such detailed curation is necessary, however, to achieve a high-quality reconstruction. It is also needed to develop an intimate familiarity with the regulatory network, which is important for prospective experimental design.

A combination of top-down and bottom-up methods is needed

There are many different data types available that relate to transcriptional regulatory networks. All these data types have information relevant to the network reconstruction process. Thus, they all need to be simultaneously reconciled. Developing methods for the reconciliation of diverse data types is one of the major challenges in network reconstruction and systems biology in general.

Six different data types that can be used in network reconstruction are illustrated in Figure 4.5. They need to be simultaneously reconciled to reconstruct a network. The end result can be depicted graphically as an interaction map. In the map shown in Figure 4.5, the transcription factors are denoted by a triangle, and the target genes by a square. The thickness of the arrows are a statistical measure of the consistency of the data types used. Ideally, all lines in the network should be thick and thus well reconciled. However, this is not typically the case, except for subnetworks or "modules." Examples of such well reconciled subnetworks are the flagellar genes in *E. coli* and those involved in nitrogen utilization in yeast [87].

4.3　Large-scale Reconstruction Efforts

Few specific transcriptional regulatory networks are well characterized. Currently, there is progress being made with a few model systems. We describe three cases here: regulation of bacterial replication, regulation of early developmental events, and regulation of a genome-scale metabolic network.

Cell cycle in *Caulobacter*

The aquatic bacterium *Caulobacter crescentus* has been extensively studied as a model organism for elucidating the cell cycle and the regulatory mechanisms that govern the precisely timed events in this cycle. *Caulobacter* divides asymmetrically into two types of progeny: a stalk cell and a motile swarmer cell. The swarmer cell typically migrates for 30–45 minutes before differentiating into a stalk cell, in which replication proceeds almost immediately. However, *Caulobacter*'s regulatory network suppresses DNA replication in swarmer cells via the global regulatory protein, CtrA, which binds to (and blocks access to) the origin of replication. Genome-scale mRNA expression studies have been used to identify over 500 genes that are regulated in a cell cycle dependent manner [118]. Figure 4.7 summarizes the key components of the *Caulobacter* cell cycle regulatory network as determined in this study. This network, which involves multiple kinases as well as the regulation of several metabolic genes, provides redundant negative feedback control over the timing of CtrA expression. More recent work has revealed how the genes in *Caulobacter* are arranged in three dimensions within the cell nucleoid as replication proceeds [237], thus expanding our thinking of this network from a "circuit" confined to two dimensions toward one that exists within more realistic three-dimensional bounds.

Early development of the sea urchin

In recent years, significant progress has been made toward a genome-scale characterization of the genetic regulatory network (GRN) responsible for controlling embryonic specification in the sea urchin, *Strongylocentrotus purpuratus*. The results from several pioneering studies have uncovered many network components and associated interactions that underlie the spatial and temporal aspects at work in the early development of this echinoderm (see Figure 4.8). The current GRN encompasses regulatory events up to 24-hour postfertilization and includes links among 50 genes including transcription factor, signaling, and their target genes [96].

Each link represented in this complex network represents the culmination of large-scale data integration from all available data, ranging from sequence-based *cis*-regulatory predictions to detailed molecular embryology. As a general requirement, experimental confirmation of each network relationship is performed via detailed expression analysis. In practice, this process involves experimentally perturbing the system, via gene knockouts for example; measuring detailed gene expression levels for all

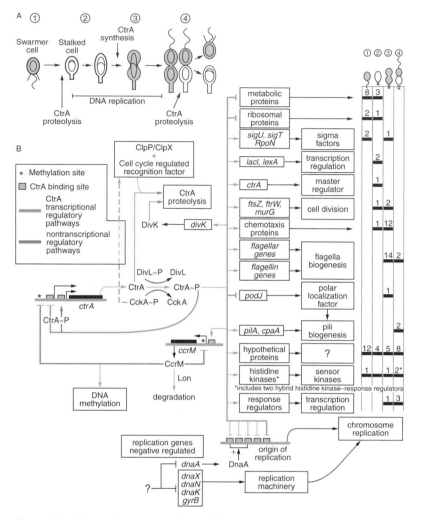

Figure 4.7: Schematic representation of the *Caulobacter crescentus* cell cycle regulatory network. Reprinted with permission from *Science* **301**:1874–1877. Copyright 2003 AAAS.

network components, and inferring network relationships from the resulting data. Confirmation of sequence-based predictions via multiple independent experimental measures cuts down on the inclusion of indirect effects, increasing the likelihood that the resulting network model reflects biological reality.

The resultant model has utility beyond simply reflecting a conceptually accurate representation of this GRN. It can also serve as an analytical tool to aid in understanding a variety of observed sea urchin developmental phenomena. For example, the current model helps explain boundary stability of differentiated cell layers, as well as the irreversibility of the embryonic

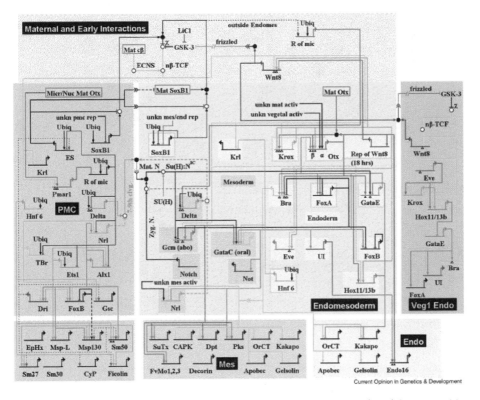

Figure 4.8: Schematic representation of the genetic regulatory network (GRN) for sea urchin endomesoderm specification. The network architecture is based on the analysis of gene expression following experimental perturbation, as well as *cis*-regulatory analyses. Reprinted from [147].

specification process [147]. Furthermore, those phenomena that are not characterized well by the model indicate poorly understood portions of the GRN and thus help direct hypothesis generation and further experiments. In any case, the success in terms of explanatory power of this still incomplete model clearly indicates the promise of this approach to modeling GRNs and will likely spur additional efforts to broaden its scope with the ultimate goal of a comprehensive and accurate model of the entire system.

Regulation of metabolism in *E. coli*

The transcriptional regulatory network in *E. coli* has been studied for the past 40 years, and as such it is one of the best characterized microbial regulatory networks. Databases, such as RegulonDB and EcoCyc, are being constructed that contain information about known regulatory interactions in this network. These known interactions were recently translated

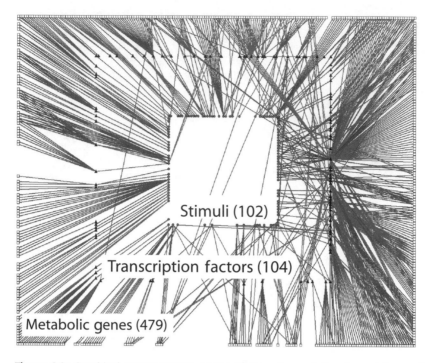

Stimuli (102)

Transcription factors (104)

Metabolic genes (479)

Figure 4.9: Graphical representation of the Boolean transcriptional regulatory network in *E. coli*. Metabolic genes are represented as squares and are arranged around the perimeter, transcription factors as triangles arranged in the middle, and stimuli as circles arranged around the center. The links or lines indicate that a Boolean rule exists relating the two objects. A link between a stimulus and a transcription factor would indicate that the stimulus affects the transcription factor activity, and a link between a transcription factor and a metabolic gene would indicate that the transcription factor regulates gene expression. Prepared by Markus Covert.

into Boolean rules or logic statements, resulting in a genome-scale model of transcriptional regulation in *E. coli*. Boolean rules were written that describe transcription factor activity as well as the conditions needed for the expression of metabolic genes. This Boolean representation of the regulatory network responds to 102 different stimuli and contains 104 transcription factors regulating 479 metabolic genes (see Figure 4.9). Once completed, this genome-scale reconstruction was used to predict growth phenotypes for single gene deletions as well as changes in expression. Discrepancies between the model predictions and experimental data led to testable hypotheses regarding both the metabolic and regulatory network [33].

Formal representation of regulatory networks
The available data on, and knowledge about, transcriptional regulatory networks can be represented in different ways. The granularity of the

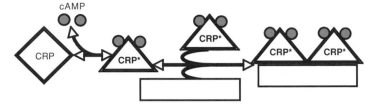

Figure 4.10: Chemical mechanisms associated with transcription factor binding to DNA. The transcription factor can exist in two conformational states, one that binds to DNA and one that does not. This mechanism is reminiscent of the way allosteric enzymes work. Courtesy of Jennie Reed.

description depends on how we intend to interrogate the data and how much detailed information we have. Thus, a spectrum of coarse-grained to detailed representations have been applied. They can be categorized as follows:

- Finding components, links, and coregulated modules by using statistical data mining methods. This represents a very high level analysis and can be used to detect patterns and screen for candidate causes and mechanisms explaining the observed effects.
- Finding causal relationships. This information can be described by directed graphs and Boolean formalism. Such logistical descriptions of regulatory networks, their states, and function have appeared [33].
- Finding reaction mechanisms. If known, they are described by chemical equations. Progress with understanding these mechanisms is being made (see Figure 4.10). The composition of the RNA polymerase and enzyme complexes is stoichiometric, and thus their formation can be described by a stoichiometric matrix.
- Finding kinetic constants. Dynamic simulations are possible if kinetic information is available [19, 52]. Given the scarcity of available numerical values for kinetic constants, such efforts are confined to small-scale networks and studies.

Given the lack of reliable reconstructions and fine-grained models of transcriptional regulatory networks, few studies have appeared that focus on their emergent properties. More such studies will eventually appear and are likely to be of key importance in systems biology, as they will help us to unravel the "logic" that cells use to manage the information on their genomes. For many, this issue is at the heart of cell and molecular biology.

4.4 Summary

> ➤ Transcriptional regulatory networks determine the expression state of a genome.

> ➤ These networks are still incompletely defined even for model organisms.

> ➤ Regulatory networks can be broken down hierarchically, into operons, regulons, and stimulons, based on the breadth of the transcriptional response.

> ➤ At the most detailed level of description, function is often described in terms of modules that are defined by investigators.

> ➤ Since many of the interactions in transcriptional regulatory networks are not known in mechanistic detail, they are often described by causal relationships.

> ➤ Once the underlying chemical mechanisms are known, the corresponding chemical equations can be described stoichiometrically.

4.5 Further Reading

Davidson, E.H., GENOMIC AND REGULATORY SYSTEMS, Academic Press, San Diego (2001).

Hasty, J., McMillen, D., Isaacs, F., and Collins, J.J., "Computational studies of gene regulatory networks: *In numero* molecular biology," *Nature Reviews Genetics*, **2**:268–279 (2001).

Herrgard, M.J., Covert, M.W., and Palsson, B.O., "Reconstruction of Microbial Transcriptional Regulatory Networks," *Current Opinion in Biotechnology*, **15**(1):70–77 (2004).

Ideker, T. and Lauffenburger, D. "Building with a scaffold: emerging strategies for high- to low-level cellular modeling," *Trends in Biotechnology*, **21**:255–262 (2003).

McAdams, H.H. and Arkin, A., "It's a noisy business! Genetic regulation at the nanomolar scale," *Trends in Genetics*," **15**:65–69 (1999).

Ptashne, M. and Gann, A., GENES AND SIGNALS, Cold Spring Harbor Press, New York (2002).

Setty, Y, Mayo, A.E., Surette, M.G. and Alon, U., "Detailed map of a cis-regulatory input function," *PNAS*, **100** 7702–7707 (2003).

Wagner, R., "Transcriptional Regulation in Prokaryotes," Oxford University Press (2000).

Wall, M.E., Hlavacek, W.S. and Savageau, M.A., "Design of Gene Circuits: lessons from bacteria," *Nature Reviews Genetics*, **5**:34–42 (2004).

Wolf, D.M., and Arkin, A.P., "Motifs, modules and games in bacteria," *Current Opinion in Microbiology*, **6**:125–124 (2003).

CHAPTER 5

Signaling Networks

Signaling networks involve the *transduction* of a "signal" from the outside to the inside of the cell. Signaling networks transmit a variety of signals from the cellular environment to the nucleus or other cellular organelles and functions. The environmental signals can be biological, such as cyto- and chemokines, or physicochemical, such as osmotic pressure or pH. Cells in multicellular organisms communicate in three principal ways:

- one cell sends a soluble signal that diffuses to the target cell,
- cells can manipulate the composition of the extracellular matrix,
- cells can communicate with very specific direct cell-to-cell mechanisms.

These fundamental modes of signal transduction rely on an underlying network of chemical reactions. Except for very few specific cases, the reconstruction of signaling networks is incomplete. However, progress and advances are currently being made toward the comprehensive reconstruction of selected signaling networks.

5.1 Basic Properties

When the cell encounters an extracellular signal (i.e., the binding of a growth factor to an extracellular receptor) a sequence of events takes place. These can be as simple as the opening of an ion channel (i.e., acetylcholine triggers the influx of calcium ions) or as complex as a highly interconnected network of protein phosphorylations. In brief, signal transduction often involves: (1) the binding of a ligand (the signaling molecule) to an extracellular receptor, (2) the subsequent phosphorylation of an intracellular enzyme, (3) amplification and passage of the signal, and (4) the

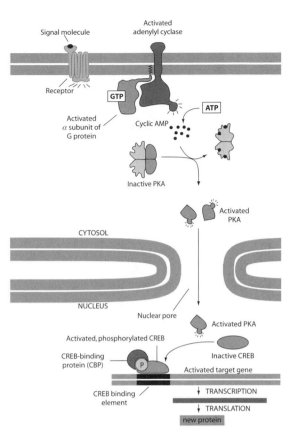

Figure 5.1: A three-level coarse-grained view of signal transduction. Redrawn from [1].

resultant change in cellular function (i.e., increase in the expression of a gene). Various classification schemas exist for the different components of signal transduction. A coarse-grained view of signaling networks shows that they have three principal parts: events around the membrane, reactions that link submembrane events to the nucleus, and events that lead to transcription (see Figure 5.1).

Signaling pathways and networks are often thought of as being different from metabolic networks. Metabolism involves the breakdown of substrate molecules for energy and redox potential production, and for the synthesis of various metabolites. The demands on metabolism are normally thought of in terms of fluxes; i.e., the cell must produce a certain amount of an amino acid to satisfy protein synthesis needs. Thus, flux maps are frequently used to describe the state of metabolism. Signaling, in contrast, conveys "information." This information is basically the transcription state of the genome. Although the result is the production of mRNA molecules (i.e., flux), it is the binding state (i.e., concentration) of the regulator sites that give the transcription state.

Some examples of signaling mechanisms are given in the following. These are just a few examples, but they serve to illustrate that signaling

1

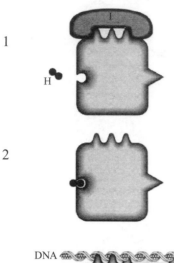

2

Figure 5.2: The basic reactions involved in steroid regulation of gene expression. Taken from [122].

3

networks involve chemical transformations that are catalyzed by enzymes. Such reactions can be described stoichiometrically.

Steroids

Perhaps one of the simplest examples of a reaction network in signal transduction is that of steroids. Sterol lipids include such hormones as cortisol, estrogen, testosterone, and calcitriol. These steroids simply cross the membrane of the target cell and then bind to an intracellular receptor that is in an inactive form due to an association with an inhibitory molecule. This binding results in the release of the inhibitory molecule from the intracellular receptor. With the steroid bound and the inhibitor released, the steroid receptor traverses the nuclear membrane and binds to its corresponding site on the DNA molecule (see Figure 5.2). This DNA binding event triggers the transcription of the target (regulated) genes.

G-protein signaling

G-protein-coupled receptors (GPCR) represent important components of signal transduction networks. For instance, this class of receptor comprises 5% of the genes in *Caenorhabditis elegans*. GPCRs consist of an extracellular domain that binds to a ligand, and another region that binds to a G-protein. The G-protein complex consists of three subunits (α, β, and γ),

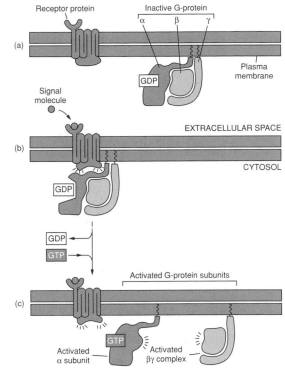

Figure 5.3: G-protein signal transduction. Adapted from [1] by permission of Garland Science/Taylor & Francis Books, Inc.

and in its inactive state is bound to guanosine diphosphate (GDP). When a ligand binds to the GPCR, the G-protein exchanges its GDP for a guanosine triphosphate (GTP). This exchange leads to the disassociation of the G-protein from the receptor and the split into a $\beta\gamma$ complex and a GTP-bound α subunit. These two components can then further relay messages to other membrane-bound molecules that transduce the signal (see Figure 5.3). The hydrolysis of the GTP-bound α subunit, replacing the GTP with GDP, leads to the reassociation of the three components of the G-protein. This inactive complex can then rebind to the GPCR. These are the basic chemical transformations that make up G-protein signaling. The system is fueled by GTP, which, like ATP, is an energy rich metabolite.

The JAK-STAT network

The JAK-STAT signaling system is an important two-step process that is involved in multiple cellular functions, including cell growth and inflammatory response. Upon binding to a cytokine, a cell surface receptor often dimerizes. The monomeric forms of the receptor are often constitutively associated with a kinase called JAK (Janus-associated kinase). In their dimerized forms, the JAKs induce phosphorylation of themselves and the receptor, activating the ligand–receptor dimer complex. This active form of the complex in turn leads to the binding of multiple proteins that

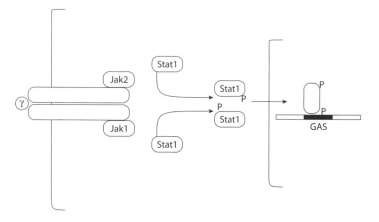

Figure 5.4: The IFN-γ-stimulated JAK-STAT pathway. Redrawn from [194].

can be further phosphorylated. One important protein that binds to the JAK–receptor complex is the STAT (signal transducers and activators of transcription) family. The STAT molecule is phosphorylated by the activated complex. The phosphorylated form of STAT then dimerizes, and this STAT dimer is then translocated to the nucleus, where it serves to activate transcription of the target genes (see Figure 5.4).

All these examples represent a set of coupled chemical reactions. Mass is balanced and thermodynamic laws are obeyed. Each of these fundamental components is a chemical transformation (Figure 3.1). The integration of the components leads to pathways, then sectors, and then whole-cell function where emergent properties surface.

Families of signaling molecules and processes
It is useful to categorize signaling processes according to shared components and to understand some basic signaling functions. One classification scheme groups signaling processes into one of 17 different families (see Table 5.1). These 17 families are further divided into three groups according to developmental processes. For example, the Wnt, receptor serine/threonine, Hedgehog, receptor tyrosine kinase (small G proteins), and Notch/Delta pathways are of particular relevance in early developmental processes of most animal cells. There may be many different isoforms of proteins that participate in each of these pathways and that provide additional specificity to a variety of signaling stimuli.

Fundamental building blocks
Some building block of signaling networks are nodes, modules, and pathways. They are detailed in the next section.

Table 5.1: The 17 intercellular signaling pathways. From [72].

Early development and later
1. Wnt pathway
2. Receptor serine/threonine kinase (TGFb) pathway
3. Hedgehog pathway
4. Receptor tyrosine kinase (small G proteins) pathway
5. Notch/Delta pathway

Mid-development and later
6. Cytokine receptor (cytoplasmic tyrosine kinases) pathway
7. IL1/Toll NFkB pathway
8. Nuclear hormone receptor pathway
9. Apoptosis pathway
10. Receptor phosphotyrosine phosphatase pathway

Adult physiology
11. Receptor guanylate cyclase pathway
12. Nitric oxide receptor pathway
13. G-protein coupled receptor (large G proteins) pathway
14. Integrin pathway
15. Cadherin pathway
16. Gap junction pathway
17. Ligand-gated cation channel pathway

Hierarchy in signaling networks

Unlike for metabolic (Figure 3.1) and transcriptional regulatory networks (Figure 4.4), well-defined hierarchical thinking has not emerged for signaling networks. The notions of modules and motifs are developing. However, a hierarchy similar to operons, regulons, and stimulons in transcriptional regulatory networks has not yet been delineated. In addition to motifs and modules, the concept of a *cross-talk* is often used to describe interactions in signaling networks. This notion based on a signal's ability to propagate beyond its "primary" channel or pathway into another. The conceptual development of hierarchy in signaling networks is likely to develop quickly, and one such development is an unbiased assessment of network properties discussed in Chapter 9.

5.2 Reconstructing Signaling Networks

Magnitude of the problem

The human genome has a repertoire of approximately 25,000 genes, each with an average of three unique transcripts [117, 236]. A human being, comprising 10^{14} cells [94] containing more than 200 different cell types [228, 94], will develop from a fertilized egg. The coordination of this developmental process and the subsequent organism's homeostatic mechanisms is achieved through signaling networks. This signaling network includes genes for 1,543 signaling receptors [236], 518 protein kinases [130], and approximately 150 protein phosphatases [65]. These components of the human signaling network result in the activation (or inhibition) of the 1,850 transcription factors [117] in the nucleus that in turn form a transcriptional regulatory network.

There is increasing evidence that the topological structure of these signaling networks is comparable to the topological structure of metabolic networks. For example, there is on average more than five metabolites that are one reaction step away from a given metabolite in a metabolic network [239]. This degree of interconnectivity is similar to that seen in the yeast signaling network in which there is on average more than five protein interactions for a given protein (as calculated from extensive data from protein–protein interaction experiments) [79]. Likewise, the structure of both networks has been described as "scale-free" in which there is a power law relationship between the network nodes and the number of links [105, 104].

These studies suggest that signaling networks are as interconnected as metabolic networks. Studies of metabolic networks have indicated that the number of functional states in a biochemical network grows much faster than the number of components [163]. This property is expected to be found in signaling networks. However, as the values here indicate, the signaling network may not be larger than some of the metabolic networks that have been reconstructed and studied to date [175]. Their analysis should therefore be possible with existing mathematical methods described in this book.

Combinatorial features

There is a large difference difference between the number of elements in a signaling network and the number of environmental stimuli that each cell type would need to respond to. However, a simple example of the power in combinatorial control demonstrates how even a very small number of elements can exert a broad spectrum of regulatory functions. The homo- and heterodimerization of only 224 proteins would provide sufficient specificity (e.g., as activating protein complexes) to control the expression of all 25,000 human genes. If a given regulatory protein were associated with multiple genes, then the number of required homo- and heterodimers for such specificity would be even less. This number is well below the estimated 1,850 transcription factors present in the human genome [117]. In fact, 1,850 transcription factors can form 1.7 million unique dimer pairs. Of course, if there are more than two components needed to induce a specific event, the combinatorial possibilities grow to correspondingly greater numbers.

Similarly, a small number of expressed receptors, used in combination, can allow for the discrimination of a very large number of environmental stimuli. For example, if we assume that 1% of the estimated 1,543 receptors in the human genome [236] are expressed in a given cell type, then that cell type could respond to 32,768 ($= 2^{15}$) different ligand combinations. These "back of the envelope" calculations emphasize that a small number of transcription factors and signaling receptor proteins operating in a combinatorial manner can allow for diversity of function in signaling networks. A recent study analyzed the repertoire of GPCRs in the human genome and identified 367 GPCRs [235]. The expression profiles of 100 GPCRs in the mouse genome for 26 different tissues indicated that most of the receptors were expressed in a variety of tissues but that each tissue had a unique profile of receptors. These results further support the existence of combinatorial control of signaling networks.

Elements of reconstruction

Signaling network reconstruction has been approached in three different ways (see Figure 3.8).

- The first approach consists of reconstructions of, preferably highly connected, *nodes*. This approach involves the delineation of all the compounds and reactions associated with a given network component (i.e., a protein, an ion, or a metabolite). For example, much work has been done with calcium that plays a key role in many signaling processes.
- The second approach consists of identifying signaling *modules*. Such modules involve grouping components that function together under certain conditions. Such grouping can be based on intuitive reasoning or on unbiased assessment of network properties (see Chapter 9). These modules allow for detailed analyses of kinetics of various concentrations and help to understand processes like feedback mechanisms. For example, much successful work has been done with the epidermal growth factor receptor and associated mitogen-activated protein (MAP) kinases [197, 248]. Analyses of other growth factor receptor signaling have also been performed [209, 164].
- The third approach of reconstructed networks involves pathways that connect signaling inputs to signaling outputs. For example, the delineation of all the steps from the binding of a growth factor to its

Table 5.2: Online sources on signaling networks.

BioCarta	http://www.biocarta.com/genes/allpathways.asp
Transpath (now commercialized)	http://transpath.gbf.de
Alliance for Cellular Signaling	http://www.signaling-gateway.org
Cell Signaling Networks Database	http://geo.nihs.go.jp/csndb
SPAD database	http://www.grt.kyushu-u.ac.jp/spad/
Science Magazine's STKE database	http://stke.sciencemag.org
Database of Quantitative Cellular Signaling (DOQCS)	http://doqcs.ncbs.res.in

receptor to the subsequent activation of a transcription factor that induces the expression of target genes. The reconstruction and analysis of the pheromone-activated MAP kinase pathway in yeast has demonstrated the utility of such an approach [227]. The reconstruction and analysis of this signaling pathway resulted in an hypothesized mechanism by which the MAP kinase Fus3p was dephosphorylated and localized at particular steps in the signaling pathway.

Level of detail in a reconstruction

An important consideration in the reconstruction of a signaling network is the desired level of detail. The level of detail can be as coarse as a delineation of associations between network components or as refined as a precise mechanistic description of the chemical reactions that occur. A reconstruction of associations can involve a description of a simple connectivity (e.g., ligand A – transcription factor B; ligand A is functionally connected to transcription factor B) or a more involved set of relationships that shows more intermediates between a signaling input and a signaling output (e.g., ligand A → protein B → protein C → transcription factor D) (e.g., [212]). Network reconstructions consisting of associations between components are amenable to multiple types of structural analyses (to be discussed). More detailed causal relationships account for cause and effect relationships (e.g., ligand A → protein B; ligand A activates protein B) (e.g., [197]). Kinetic relationships build off of these causal relationships, assigning scaling factors and time constants between different properties of interest. At an even more refined level of detail are mechanistic reconstructions. These reconstructions account for stoichiometric relationships between signaling components (e.g., ligand A binds to receptor B and receptor B then dimerizes) and thus can be represented with stoichiometric matrices. This level of detail allows for an accounting of all network components (e.g., ATP and receptor protein synthesis) necessary to drive a signal from stimulus to response.

Data sources for reconstruction process

High-throughput techniques to elucidate the mechanisms that connect an extracellular signaling stimulus to the level of control by transcription factors are still in their infancy [260]. Despite their shortcomings, such technologies are leading to the characterization of intracellular signaling mechanisms at a large scale. These developing technologies can be grouped into two categories: first, biochemical techniques and expression systems for characterizing protein–protein interactions; and second, assays for piecing together functional properties.

1. *Characterizing interactions.* Perhaps the most widely used technique for deciphering protein–protein interactions involves yeast two-hybrid assays. However, to date, there is only a small degree of congruence between different data sets. False-positive results for protein-protein interactions in yeast two-hybrid experiments occur in part because spatially or temporally segregated proteins would not interact *in vivo*.

Multiprotein complexes in *Saccharomyces cerevisiae* have been characterized using mass spectrometric approaches [70, 89]. Additional biochemical techniques for investigating intracellular signaling networks are developing, including isotope-coded affinity tags, stable isotope labeling by amino acids in cell culture, Src-homology-2 profiling, and target-assisted iterative screening [214]. Although these approaches are only beginning to be systematically applied at a large scale, the initial results are promising [16].

2. *Assays for functional properties.* Four approaches are highlighted here. First, perturbation analysis monitors genome-wide changes in gene expression after disrupting specific components of a network and has been used to refine models of the yeast galactose-utilization pathway [100]. Second, knockdown strategies (RNAi) have been used to elucidate the components of the Hedgehog signaling pathway in *Drosophila melanogaster* [126], and the interactions between deubiquitylating enzymes and the IKK complex involved in NF-kB signaling [20]. This approach will certainly develop into a powerful tool for deciphering larger cellular signaling networks at a genome scale [18, 259]. Third, protein arrays in development generate high-throughput data on protein presence and activity [145]. Fourth, fluorescence imaging technologies are generating data regarding protein localization and the dynamics of signaling processes [137, 167]. For example, a GFP-fusion genomic DNA library of *Schizosaccharomyces pombe* was created by fusing fragments of the

S. pombe genome to the GFP gene. *S. pombe* cells were then transformed with this plasmid library, and the localization of proteins to 11 distinct compartments was evaluated [46].

Imaging technology allows for a global analysis of protein localization in *S. cerevisiae* and discriminated between 22 different cellular locales. It was also used to assign a cellular compartment location for 70% of proteins for which the location was not previously known [98]. Furthermore, fluorescence resonance energy transfer (FRET) is used to decipher specific signaling mechanisms because it can indicate molecular proximity. For example, FRET has been used to study membrane-extracellular signaling events and showed that 14% of epidemal growth factor (EGF) receptors in A431 cells were oligomerized before growth factor binding [132]. It has also been used to study membrane-associated signaling mechanisms (e.g., activation of heterotrimeric G-protein complexes might involve rearrangement rather than dissociation [22]), and intracellular signaling events (e.g., the phosphorylation states of insulin–receptor substrates [189]).

Integration of data types

Each of these techniques has distinct advantages and disadvantages. There is thus a growing need to integrate a variety of data sources to most accurately reconstruct a signaling network. Initial efforts to integrate disparate data sources have been successful. For example, yeast two-hybrid protein–protein interaction experiments, RNAi phenotyping, and gene expression arrays were used for *Caenorhabditis elegans* germline to generate systems-level hypothesis including the tendency of essential proteins to interact with each other [240]. The integration of experimental data from experimental sources in the context of a mathematical model has also been performed [85]. Regulatory interactions determined from gene expression arrays were evaluated for consistency with reconstructed regulatory networks of *E. coli* and *S. cerevisiae*. Novel hypotheses were generated from this data integration study. For example, expression arrays and literature-based reconstructions are generally less consistent for repressor than activator regulatory proteins. These systems-level descriptions require the integration of multiple types of data since no single experimental protocol can accurately characterize all necessary parameters for a systems-level biological description.

Large-scale reconstruction efforts

To date, most work on reconstructing signaling networks has been limited to analyses at a small scale, i.e., analyzing the dynamics of a particular receptor–ligand complex. However, with the availability of large-scale data

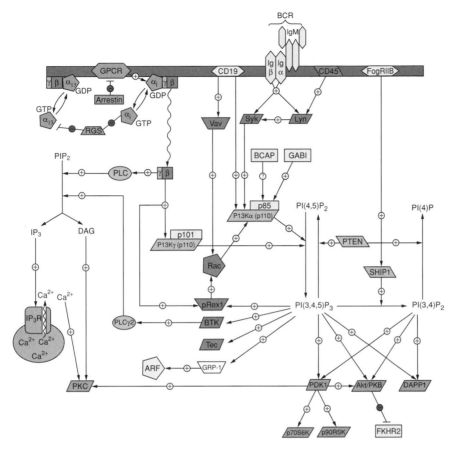

Figure 5.5: An example of a complex signaling network. Taken from http://www.afcs.org.

sets, the scope of such reconstructions is growing. The Alliance for Cellular Signaling [74] has focused resources on elucidating signaling mechanisms in the human B-cell and cardiac myocyte, and more recently a human macrophage cell line. The Cell Migration (www.cellmigration.org) and LIPID MAPS (http://www.lipidmaps.org/) consortiums have also begun work on elucidating components of signaling networks.

The descriptions of the signaling mechanisms described earlier only hint at the intricate and complex networks that are formed from the interactions of all components in signaling networks (see Figure 5.5). Large-scale efforts are required (see http://www.afcs.org for a description of the multi-institutional effort in the Alliance for Cellular Signaling) to decipher all the players and relationships in the tremendous reaction networks that exist in living cells. With the significant effort currently being devoted to signaling, we may expect to see cell and tissue-scale networks emerge over the coming years.

5.3   Summary

➤ Signal transduction involves the transmission of extracellular signals into the nucleus of the cell, leading to changes in gene expression.

➤ There are three different modes of cellular communication: soluble signaling, extracellular matrix–cell, and direct cell–cell.

➤ Signaling pathways involve three basic steps: the formulation of a membrane complex, a series of reactions leading to the nucleus, and changed activity of transcription factors.

➤ Signaling networks have combinatorial properties.

➤ Few signaling pathways have been extensively reconstructed.

➤ Reconstruction of signaling pathways involves the integration and use of multiple data types (many of which are similar to those used to reconstruct transcriptional regulatory networks).

5.4   Further Reading

Gompets, B.D., Kramer, I.M., and Tatham, P.E.R., SIGNAL TRANSDUCTION, Academic Press, San Diego (2002).

Manning, G., Whyte, D.B., Martinez, R., Hunter, T., and Sudarsanam, S., "The protein kinase complement of the human genome," *Science*, **298**:1912–1934 (2002).

Papin, J.A., Hunter, T., Palsson, B.O., and Subramaniam, S., "Reconstruction of cellular signaling networks and analysis of their properties," *Nature Reviews Molecular Cell Biology*, **6**:99–111 (2005).

Tyson, J.J., Novak, B., Odell, G.M., Chen, K., and Thron, C.D., "Chemical kinetic theory: Understanding cell cycle regulation," *Trends in Biochemical Sciences*, **21**:89–96 (1996).

PART TWO

Mathematical Representation of Reconstructed Networks

The set of chemical reactions that comprise a network can be represented as a set of chemical equations. Embedded in these chemical equations is information about reaction stoichiometry. All this stoichiometric information can be represented in a matrix form; the stoichiometric matrix, denoted by **S**. Associated with this matrix is additional information about enzyme complex formation, transcript levels, open reading frames, and protein localization. Therefore, once assembled, the stoichiometric matrix represents a biochemically, genetically, and genomically (BIGG) structured database. This database structure represents an interface between high-throughput data and *in silico* analysis (see Figure 1.4). It allows high-throughput data (often called *content*) to be put into context. The stoichiometric matrix is the starting point for various mathematical analysis used to determine network properties. Part II of this text will summarize the basic properties of the stoichiometric matrix. Since it is a mathematical object, the treatment is necessarily mathematical. However, **S** represents biochemistry. We will thus relate the mathematical properties of **S** to the biochemical and biological properties that it fundamentally represents.

CHAPTER 6

Basic Features of the Stoichiometric Matrix

The stoichiometric matrix is formed from the stoichiometric coefficients of the reactions that comprise a reaction network. This matrix is organized such that every column corresponds to a reaction and every row corresponds to a compound (recall Figure 1.5). The entries in the matrix are stoichiometric coefficients, which are integers. Each column that describes a reaction is constrained by the rules of chemistry, such as elemental balancing. Every row thus describes the reactions in which that compound participates and therefore how the reactions are interconnected. The stoichiometric matrix transforms the flux vector (that contains the reaction rates) into a vector that contains the time derivatives of the concentrations. The stoichiometric matrix thus contains chemical *and* network information. These basic properties of the stoichiometric matrix are described in this chapter.

6.1 S as a Linear Transformation

Mathematically, the stoichiometric matrix $\mathbf{S}$ is a *linear transformation* (Figure 6.1) of the flux vector

$$\mathbf{v} = (v_1, v_2, \ldots, v_n) \tag{6.1}$$

to a vector of time derivatives of the concentration vector

$$\mathbf{x} = (x_1, x_2, \ldots, x_m) \tag{6.2}$$

as

$$\frac{d\mathbf{x}}{dt} = \mathbf{S}\mathbf{v} \tag{6.3}$$

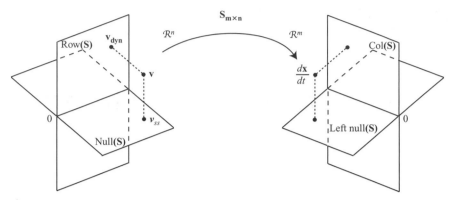

Figure 6.1: The stoichiometric matrix as a linear transformation. The four fundamental subspaces of **S** are shown. Prepared by Iman Famili.

The reader may also be familiar with other notations of time derivatives

$$\frac{d\mathbf{x}}{dt} = \mathbf{x}' = \dot{\mathbf{x}} \tag{6.4}$$

This perhaps makes it more clear that the $d\mathbf{x}/dt$ is a vector and that $\dot{\mathbf{x}} = \mathbf{Sv}$ is a linear transformation.

Dynamic mass balances

Equation 6.3 represents the fundamental equation of the *dynamic mass balances* that characterizes all functional states of a reconstructed biochemical reaction network. Each individual equation in the set

$$\frac{dx_i}{dt} = \sum_k s_{ik} v_k \tag{6.5}$$

represents a summation of all fluxes v_k that form compound x_i and those that degrade it.

Dimensions

There are m metabolites (x_i) found in the network and n reactions (v_i). Thus,

$$\dim(\mathbf{x}) = m, \quad \dim(\mathbf{v}) = n, \quad \dim(\mathbf{S}) = m{\times}n \tag{6.6}$$

For a typical biological network there are more reactions than compounds, or $n > m$. The matrix **S** may not be full rank, and therefore $\mathrm{Rank}(\mathbf{S}) = r < m$.

The four fundamental subspaces

There are four fundamental subspaces associated with a matrix. The four fundamental subspaces of **S**, shown in Figure 6.1, have important roles in the analysis of biochemical reaction networks, as detailed in the following chapters. The vector produced by a linear transformation is in two orthogonal spaces (the column and left null spaces), called the *domain*, and the vector being mapped is also in two orthogonal spaces (the row and null spaces), called the *codomain* or the *range* of the transformation.

The column and left null spaces

The time derivative is in the column space of **S** (denoted by Col(**S**)), as can be seen from the expansion of **Sv**:

$$\frac{d\mathbf{x}}{dt} = \mathbf{s}_1 v_1 + \mathbf{s}_2 v_2 + \cdots + \mathbf{s}_n v_n \tag{6.7}$$

where $\mathbf{s}_i$ are the *reaction vectors* that form the columns of **S**. Col(**S**) is therefore spanned by the reaction vectors $\mathbf{s}_i$. The reaction vectors are structural features of the network and are fixed. However, the fluxes v_i are scalar quantities and represent the flux through reaction i. The fluxes are variables. We do note that each flux has a maximal value, $v_i \leq v_{i,max}$, and this limits the size of the time derivatives. Thus, only a portion of the column space is explored, that is, we can cap the size of the column space of **S**. The vectors in the left null space ($\mathbf{l}_i$) of **S** are orthogonal to the column space, that is, $\langle \mathbf{l}_j \cdot \mathbf{s}_i \rangle = 0$. The vectors $\mathbf{l}_i$ represent a mass conservation, (see Chapter 10).

The row and null spaces

The flux vector can be decomposed into a dynamic component and a steady-state component:

$$\mathbf{v} = \mathbf{v}_{\text{dyn}} + \mathbf{v}_{\text{ss}} \tag{6.8}$$

The steady state component satisfies

$$\mathbf{S}\mathbf{v}_{\text{ss}} = 0 \tag{6.9}$$

and $\mathbf{v}_{\text{ss}}$ is thus in the null space of **S** (see Chapter 9). The dynamic component of the flux vector, $\mathbf{v}_{\text{dyn}}$, is orthogonal to the null space and consequently it is in the row space of **S**. Each pair of subspaces in the domain and codomain of the dynamic mass balance equation therefore form orthogonal sets to each other, and their dimensions sum up to the dimension of their corresponding vectors, that is, dim(Null(**S**) + dim(Row(**S**)) = n and dim(Left null(**S**)) + dim(Col(**S**)) = m.

These are introductory observations about **S** and its fundamental subspaces. In Chapters 8 through 11, we will study the individual fundamental subspaces in more detail.

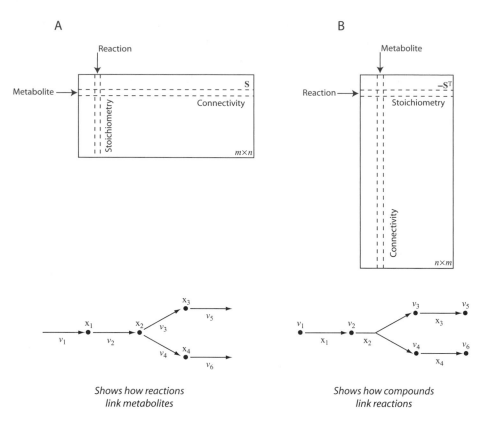

Figure 6.2: The structure of the stoichiometric matrix and how it corresponds to a map. Both the regular (reaction) and transpose (compound) maps are shown in (A) and (B), respectively. Prepared by Iman Famili.

6.2   **S** as a Connectivity Matrix

In the stoichiometric matrix, each column represents a reaction and each row represents a compound (Figure 6.2A). **S** is a *connectivity matrix*, and it represents a network. This network is represented with a *map*. Each *node* in the map corresponds to a row in the matrix, and each column corresponds to a *link* in the map. Therefore, **S** represents a map where a compound is a node and the reactions connect (link) the compounds. This map is the *reaction map* and is the standard way of viewing metabolic reactions and pathways in biochemistry textbooks.

The negative of the transpose of the stoichiometric matrix, $-\mathbf{S}^{\mathrm{T}}$, also represents a map (Figure 6.2B), which we will call the *compound map*. The map that $-\mathbf{S}^{\mathrm{T}}$ represents has the reactions (now the rows in $-\mathbf{S}^{\mathrm{T}}$) as the nodes in the network and the compounds (now the columns of $-\mathbf{S}^{\mathrm{T}}$) as the connections, or the links. This representation of a biochemical reaction network is unconventional but useful in many circumstances.

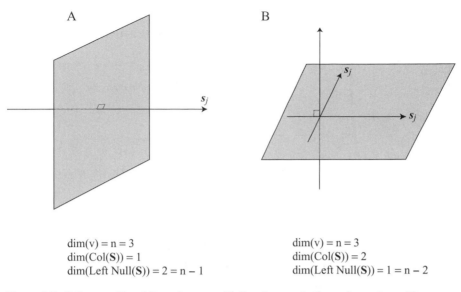

dim(v) = n = 3
dim(Col(**S**)) = 1
dim(Left Null(**S**)) = 2 = n − 1

dim(v) = n = 3
dim(Col(**S**)) = 2
dim(Left Null(**S**)) = 1 = n − 2

Figure 6.3: Orthogonality of the column and left null space in three dimensions: (A) a one-dimensional and (B) a two-dimensional column space.

It is worth examining the columns ($\mathbf{s}_i$) and rows of **S** a bit more closely. Let us examine a reaction:

$$A + B \xrightarrow{v_i} C + D \tag{6.10}$$

with the corresponding column of **S**, $\mathbf{s}_i = (-1, -1, 1, 1)^{\mathrm{T}}$. This vector is in the column space of **S**. Moving along this vector is like carrying out this reaction. Note that motion along this vector will conserve the sum $A + B + C + D$. Thus, a column in **S** represents a "tie" between the compounds participating in a particular reaction. If these compounds participate in other reactions, there will be interactions between the motions along the columns of **S**. These vectors, $\mathbf{s}_i$, span the column space of **S** and thus give a conceptually useful basis for the column space of **S**.

The orthogonality of the column and left null space can be represented schematically for a three-dimensional case (Figure 6.3). If there is one independent reaction vector $\mathbf{s}_i$, there is a two-dimensional subspace orthogonal to this single reaction where the network's metabolite mass is conserved. If the system has two independent reactions, there is only a one-dimensional left null space. Thus, the higher the number of independent reaction vectors, the smaller the orthogonal left null space. The higher the number of independent reactions, the fewer conservation quantities exist.

Similar comments follow for the rows of **S** (or the columns of $-\mathbf{S}^{\mathrm{T}}$). A column in $-\mathbf{S}^{\mathrm{T}}$ will tie together, or connect, all the reactions in which a metabolite participates. Note that the columns of **S** create a "hard" connection between the metabolites, since a reaction will simultaneously use

$$
\begin{pmatrix}
\begin{array}{ccc}
v_1 & v_2 & v_3 \\
0 & 0 & 1 \\
0 & -1 & -1 \\
0 & -1 & 0 \\
-1 & 1 & 1 \\
0 & 0 & -1 \\
1 & 0 & 0
\end{array}
\end{pmatrix}
\begin{array}{l}
A \\
C \\
P \\
CP \\
AP \\
PC
\end{array}
$$

$$CP \xleftrightarrow{v_1} PC \qquad \textit{Reversible conversion}$$

$$C + P \xleftrightarrow{v_2} CP \qquad \textit{Bimolecular association}$$

$$C + AP \xleftrightarrow{v_3} CP + A \quad \textit{Cofactor-coupled reaction}$$

Figure 6.4: The columns of the stoichiometric matrix that correspond to elementary chemical transformations.

and produce the participating compounds. Conversely, the connectivities created between the reactions are "soft," since the reactions in which a compound participates can have varying flux levels that may not have fixed ratios. These ratios are determined by the kinetic properties of the reactions.

6.3 Elementary Biochemical Reactions

There is a limited number of elementary types of biochemical reactions that take place in cells. These fall into three categories. In the following examples derived for metabolic transformations, we used C to denote a primary metabolite, P as a phosphate group, and A as a cofactor such as the adenosine moiety in AMP, ADP, and ATP. The columns, s_i, that correspond to these elementary transformations are shown in Figure 6.4.

Reversible conversion
Transformation between two compounds comprising the same two chemical moieties C and P can be written as

$$CP \rightleftharpoons PC \tag{6.11}$$

representing two elementary reactions (forward and backward). Although such reversible conversions are often used to generically describe reactions, they can only represent simple chemical rearrangement of the molecule without any change in its elemental composition. Isomerases catalyze such reactions. The stoichiometric matrix that describes this reaction is

$$\mathbf{S} = \begin{pmatrix} -1 & 1 \\ 1 & -1 \end{pmatrix} \tag{6.12}$$

where the first column of the matrix represents the forward reaction and the second column the reverse reaction. The first row represents CP, and the second row PC. Under certain circumstances, one may wish to combine the two elementary reactions into a net reaction that can take on positive or negative values.

Bimolecular association

Many biochemical reactions involve the combination of two moieties, C and P, to form a new compound.

$$C + P \rightleftharpoons CP \tag{6.13}$$

Sometimes, such reactions may not involve covalent bonds but a series of hydrogen bonds to form a complex, such as the dimerization of two protein molecules or the initial binding of a substrate to an active site on an enzyme molecule. The stoichiometric matrix that describes a bimolecular association is

$$\mathbf{S} = \begin{pmatrix} -1 & 1 \\ -1 & 1 \\ 1 & -1 \end{pmatrix} \tag{6.14}$$

where the rows represent C, P, and CP, respectively, and the columns represent the forward and reverse elementary reactions.

A cofactor-coupled reaction

A frequent reaction in biochemical reaction networks is one in which one compound (AP) donates a moiety (P) to another compound (C):

$$C + AP \rightleftharpoons CP + A \tag{6.15}$$

In reality, such reactions have an intermediate and can be decomposed into two bimolecular association reactions. The stoichiometric matrix that describes the cofactor-coupled (or exchange) reaction is

$$\mathbf{S} = \begin{pmatrix} -1 & 1 \\ -1 & 1 \\ 1 & -1 \\ 1 & -1 \end{pmatrix} \tag{6.16}$$

where the rows represent C, AP, CP and A, respectively, and the columns represent the forward and reverse reactions. The word *cofactor* is used synonymously with *carrier*.

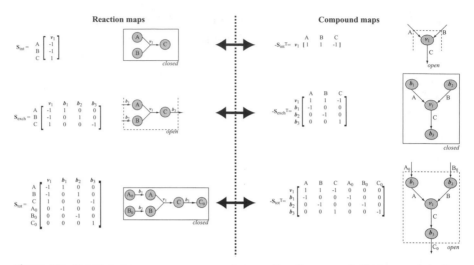

Figure 6.5: Reaction maps versus compound maps. Reaction maps (left) show metabolites as nodes and reactions as directed edges. The reaction map includes both the internal and exchange fluxes, if present. In contrast, compound maps of the same systems (right) show the reactions as nodes and metabolites as directed edges. A system boundary that allows for the exchange of the internal nodes is open on a reaction map. The compound map of an open reaction map is closed, and vice versa, as is shown by changing the network from (A) to (B) and to (C). Figure prepared by Iman Famili.

6.4   Linear and Nonlinear Maps

The topological structure of the maps formed by connectivity matrices are very important in determining the properties of the network. The topological properties of maps can be linear and nonlinear (Figure 6.4). Some simple examples of reaction and compound maps are shown in Figure 6.5.

Linear maps are made up of links that have only one input and one output. Thus, the columns of $\mathbf{S}$ will only have two entries, corresponding to the two nodes (metabolites) that the link (reaction) connects. Similarly for $\mathbf{S}^T$, if only one compound links two reactions, the map is linear. Although frequently used for illustrative purposes, the occurrence of such links in biological reaction networks is rare.

Nonlinear maps are made up of links with more than one input or more than one output (Figure 6.4). The number of compounds that participate in a reaction can be found by adding up the nonzero elements in the corresponding column of $\mathbf{S}$. In genome-scale metabolic models, the most common number of metabolites participating in a reaction is 4, as in reaction 6.15. Thus, metabolic cofactors create nonlinearity in the map of $\mathbf{S}$. Metabolites that participate in more than two reactions create a nonlinearity in the map of $\mathbf{S}^T$. The participation number of a metabolite in genome-scale models can be as high as 150 (for ATP) but is 2 for most compounds found in the cell [50]. The metabolites that participate in many reactions thus create strong nonlinearities in the compound map. Again, the cofactors lead to strong nonlinear topological features of metabolic networks. However, note the difference in the maps: in the reaction map, the nonlinear links are hard (i.e., two molecules must come together to produce a reaction), whereas in the compound map, a molecule can go through one reaction or another.

6.5 The Elemental Matrix

The chemical reactions that form the columns in $\mathbf{S}$ have the basic rules of chemical transformation associated with them. There are conservation quantities (such as elements and charge) and there are nonconserved quantities (such as osmotic pressure and free energy) associated with chemical transformations. These properties must be accounted for in the construction of a biochemically meaningful stoichiometric matrix.

Every compound in the reaction network comprises chemical elements. The elemental matrix $\mathbf{E}$ gives the composition of all the compounds considered in a network. A column of $\mathbf{E}$ corresponds to a compound, $\mathbf{e}_i$, and the rows correspond to the elements, typically carbon, oxygen, nitrogen, hydrogen, phosphorous, and sulfur. It is important to note that the elemental composition of a molecule does not uniquely specify its chemical structure. For instance, glucose and fructose have the same elemental structure. The elemental composition of common metabolites is shown in Table 6.1.

Example: Consider the simple chemical reaction

$$2H_2 + O_2 \rightarrow 2H_2O \tag{6.17}$$

that involves only two elements, oxygen and hydrogen. The elemental matrix for this chemical reaction is

$$\mathbf{E} = \begin{pmatrix} 0 & 2 & 1 \\ 2 & 0 & 2 \end{pmatrix} \tag{6.18}$$

where the first row corresponds to oxygen and the second row to hydrogen. The columns correspond to the compounds, in this case ordered as H_2, O_2, and H_2O.

Table 6.1: The elemental composition of some common metabolites. Adapted from [144].

Compound	Elemental composition	Compound	Elemental composition
Glucose	$C_6H_{12}O_6$	Alanine	$C_3H_7NO_2$
Glucose-6-phosphate	$C_6H_{11}O_9P$	Arginine	$C_6H_{14}N_4O_2$
Fructose-6-phosphate	$C_6H_{11}O_9P$	Asparagine	$C_4H_8N_2O_3$
Fructose-1, 6-phosphate	$C_6H_{10}O_{12}P_2$	Cysteine	$C_3H_7O_2NS$
Dihydroxyacetone phosphate	$C_3H_5O_6P$	Glutamic acid	$C_5H_9NO_4$
Glyceraldehyde-3-phosphate	$C_3H_5O_6P$	Glycine	$C_2H_5NO_2$
1,3-Diphosphoglycerate	$C_3H_4O_{10}P_2$	Leucine	$C_6H_{13}NO_2$
2,3-Diphosphoglycerate	$C_3H_3P_2O_{10}$	Isoleucine	$C_6H_{13}NO_2$
3-Phosphoglycerate	$C_3H_4O_7P$	Lysine	$C_6H_{14}N_2O_2$
2-Phosphoglycerate	$C_3H_4O_7P$	Histidine	$C_6H_9N_3O_2$
Phosphoenolpyruvate	$C_3H_2O_6P$	Phenylalanine	$C_9H_{11}NO_2$
Pyruvate	$C_3H_3O_3$	Proline	$C_5H_9NO_2$
Lactate	$C_3H_5O_3$	Serine	$C_3H_7NO_3$
6-Phosphogluco-lactone	$C_6H_9O_9P$	Threonine	$C_4H_9NO_3$
6-Phosphogluconate	$C_6H_{10}O_{10}P$	Tryptophane	$C_{11}H_{12}N_2O_2$
Ribulose-5-phosphate	$C_5H_9O_8P$	Tyrosine	$C_9H_{11}NO_3$
Ribulose-5-phosphate	$C_5H_9O_8P$	Valine	$C_5H_{11}NO_2$
Xylulose-5-phosphate	$C_5H_9O_8P$	Methionine	$C_5H_{11}O_2NS$
Ribose-5-phosphate	$C_5H_9O_8P$	Sedoheptulose 7-phosphate	$C_7H_{13}O_{10}P$
Erythrose-4-phosphate	$C_4H_7O_7P$	5-Phosphoribosyl 1-pyrophosphate	$C_5H_8O_{14}P_3$
Inosine monophosphate	$C_{10}N_4H_{12}O_8P$	Ribose-1-phosphate	$C_5H_9O_8P$
Hypoxanthine	$C_5N_4H_4O$	Inosine	$C_{10}H_{12}N_4O_5$

Conserved quantities

A chemical reaction cannot create or destroy elements. Thus, the inner product of the rows, e_i, in the elemental matrix and the reaction vectors, s_j, must be zero, or

$$\langle e_i \cdot s_j \rangle = 0 \qquad (6.19)$$

for all the elements found in the compounds that participate in the reaction. This inner product simply adds up an element on each side of the reaction. Since the stoichiometric coefficients are negative for the reactants (the compounds that disappear in the reaction) and positive for the products (the compounds that appear in the reaction), this sum is zero. The number of atoms of an element on each side of the reaction is the same. For the elemental matrix in equation 6.18 and the reaction vector $\mathbf{s}_i = (-2, -1, 2)^{\mathrm{T}}$ we see that

$$\langle (0, 2, 1) \cdot (-2, -1, 2)^{\mathrm{T}} \rangle = 0 \quad \text{and} \quad \langle (2, 0, 2) \cdot (-2, -1, 2)^{\mathrm{T}} \rangle = 0 \qquad (6.20)$$

All elemental balancing equations taken together lead to

$$\mathbf{ES} = \mathbf{0} \qquad (6.21)$$

Although not shown here, the same must be true of compound charge, since it is balanced during a chemical reaction.

Isotopomers are often used to trace the flow of atoms, typically carbon, through a metabolic network. In this case, the carbon atoms are not identical since some carbon atoms in the substrate may be of atomic mass 13 and not the usual 12. The fate of these particular carbon atoms can be traced using particular conservation rules [112, 195, 252, 247].

Nonconserved quantities

Other physicochemical properties may not be conserved during a chemical reaction. Such properties of the molecules can be represented as an appended row to the elemental matrix. For instance, all the osmotic coefficients for reactants and products can be listed. If they do not sum to zero, the osmotic pressure will not be balanced as the reaction takes place. In other words, osmolality of the solution can be increased or decreased as a reaction proceeds. Gibbs free energy is another important quantity that changes with chemical reaction. When summed, the Gibbs free energy change of a reaction needs to be negative for a reaction to proceed forward at a significant rate. Note that nonconserved properties do lie in the row space of $\mathbf{S}$ since the summation of the rows of $\mathbf{S}$ for these properties is nonzero.

Compounds as points in the elemental space

All compounds contain a finite number of elements, described with integral numbers. Thus, any compound can be represented in a space where the axes correspond to the elements (see Figure 6.6). All feasible combinations of the elements represented by the axes of the space can be represented as point in the elemental space. For instance, the water molecule is in the $(0, 2, 1)$ point in a three-dimensional space formed by carbon, hydrogen, and oxygen. There are six elements (carbon, hydrogen, oxygen, nitrogen, phosphate, and sulfur) that make up most biochemical molecules.

Reaction vectors as connections between these points

The reaction vectors connect points in the elemental space (Figure 6.7). Isomerization like the one indicated in equation 6.11 is simply a point in the elemental space, since the elemental composition of the compound

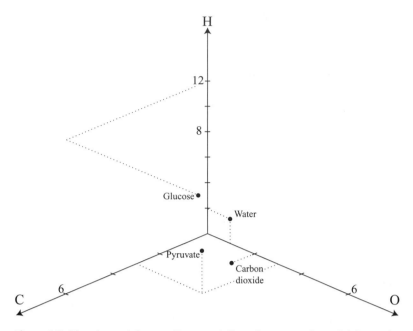

Figure 6.6: The elemental space. Representation of compounds containing carbon, hydrogen, and oxygen in a three-dimensional space. The coordinates are glucose (6, 12, 6), pyruvate (3, 3, 3), water (0, 2, 1), and carbon dioxide (1, 0, 2).

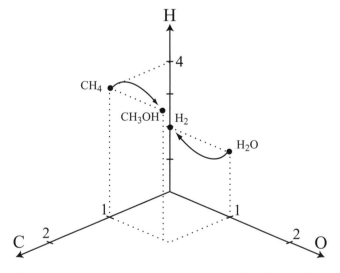

Figure 6.7: Reactions as co-connected dots in the elemental space. The reactant and products of the reaction $H—CH_3 + H—OH \rightarrow CH_3—OH + H_2$ are shown. The arrows indicated are meant to connect all the dots at the same time.

does not change. Since most reactions involve more than one reactant and more than one product, the ties are not simple lines. If the location of a compound along an axis is normalized to its stoichiometric coefficient in the reaction (no change if it is unity), then the ties that the reactions represent are orthogonal to the axes since the number of elements must be conserved.

Chemical moieties

Chemical moieties that do not change in the network can be represented as a group that displays the combination of the elements it comprises. For instance, in the chemical reaction

$$H-CH_3 + H-OH \rightarrow CH_3-OH + H_2 \tag{6.22}$$

we can consider the CH_3 and OH groups as moieties that are intact in this reaction. Note that this reaction is of the type shown in equation 6.15, where OH and H are being exchanged between CH_3 and H.

The elemental matrix for this reaction is

	CH_4	H_2O	CH_3OH	H_2
C	1	0	1	0
H	4	2	4	2
O	0	1	1	0

It can also be written with the invariant moieties as rows

	CH_4	H_2O	CH_3OH	H_2
C	0	0	0	0
H	1	1	0	2
O	0	0	0	0
CH_3	1	0	1	0
OH	0	1	1	0

Note that the rows for carbon and oxygen now have all zero entries.

Metabolic carrier molecules as conserved moieties

The consideration of conserved chemical moieties is useful in describing transformations in biochemical reaction networks. Many cofactors, such as ATP and NADH, do not change except for a phosphate group (a chemical moiety) and a redox equivalent in the form of a hydrogen ion.

Phosphorylated adenosines are carriers of high-energy bonds. We can think of ATP as AMP-P-P, and ADP as AMP-P. When the high-energy

Table 6.2: The elemental composition of common chemical moieties in metabolism. Adapted from [144].

Compound	Elemental composition	Compound	Elemental composition
Adenine	$C_5H_5N_5$	Adenosine	$C_{10}H_{13}N_5O_4$
Adenosine monophosphate	$C_{10}N_5H_{13}O_7P$	Adenosine diphosphate	$C_{10}N_5H_{13}O_{10}P_2$
Adenosine triphosphate	$C_{10}N_5H_{13}O_{13}P_3$	Nicotinamide adenine dinucleotide	$C_{21}H_{27}N_7O_{14}P_2$
Nicotinamide adenine	$C_{21}H_{28}N_7O_{14}P_2$		
Nicotinamide adenine dinucleotide phosphate	$C_{21}H_{27}N_7O_{17}P_3$	Nicotinamide adenine dinucleotide phosphate	$C_{21}H_{28}N_7O_{17}P_3$
Hydrogen ion	H	Inorganic phosphate	HPO_4
Ammonia	NH_3	Carbon dioxide	CO_2
Water	H_2O		

phosphate bond is transferred between these compounds, the AMP portion remains invariant. In the case of the redox carrier NAD$^+$ and NADH, the core cofactor molecule is conserved. However, in this case we also have to conserve compound charge, and so H$^+$ becomes an important player in redox conservation quantities (see Chapter 10 for details).

There are several types of carrier molecules in metabolism (see Table 3.1). Any cyclic process has a chemical moiety at its core. For example, in the TCA cycle there is a conserved C_4 moiety that cycles around in this pathway (e.g., see Table 10.8). Similar conserved moieties exist in other cyclic pathways. The elemental composition of some common chemical moieties are shown in Table 6.2. These moieties tend to be conserved on faster time scales.

Protein molecules as conserved moieties

In enzyme-catalyzed reactions, the backbone of the enzyme molecule can be considered as an invariant moiety since the enzyme molecule is recovered intact after the reaction has taken place. For instance, consider the glucosephosphate isomerase (PGI) in glycolysis. It catalyzes the reaction

$$G6P + E \rightleftharpoons E\text{-}G6P \rightleftharpoons E + F6P$$

The elemental composition of G6P and F6P is the same (Table 6.1), and the elemental composition of this enzyme itself does not change during this reaction. The basic enzyme can be considered as a moiety. In reaction

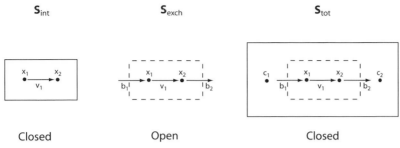

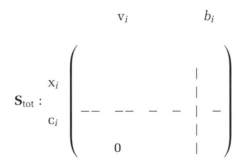

Figure 6.8: Schematic illustration of open and closed metabolic networks.

networks of signaling pathways and regulation of DNA transcription, one can identify similar conserved moieties.

6.6   Open and Closed Networks

The boundaries around a network can be drawn in different ways (see Figure 6.8). In defining a network a *systems boundary* is drawn. The reactions are then partitioned into internal and exchange reactions. Exchange fluxes are denoted by b_i and internal fluxes by v_i. Similarly, the concentration vector is partitioned into internal (x_i) and external concentrations (c_i). There are several different versions of **S** depending on what is encompassed by a network.

The total stoichiometric matrix
The most general form of **S** is

$$
\mathbf{S}_{tot}: \begin{matrix} & v_i & b_i \\ x_i & & \\ & \left(\begin{array}{ccc|c} & & & | \\ -- & -- & - - & | & - \\ & & & | \\ 0 & & & | \end{array} \right) \\ c_i & & \end{matrix}
$$

where the dashed lines show the partitioning of the internal elements in the matrix. This form accounts for the internal reactions (v_i), the exchange reactions (b_i), the internal compounds (x_i), and the external compounds (c_i).

The exchange stoichiometric matrix

If we do not consider the external compounds, c_i, we have

$$
\mathbf{S}_{\text{exch}} : x_i \quad
\begin{array}{cc}
v_i & b_i \\
\left(
\begin{array}{cc}
 & \vdots \\
 & \vdots \\
 & \vdots \\
 & \vdots \\
 & \vdots
\end{array}
\right)
\end{array}
$$

which only contains the internal fluxes and the exchange fluxes with the environment. This form of the matrix is frequently used in pathway analysis of a network (see Chapter 9).

The internal stoichiometric matrix

If we consider the cell as a closed system, we can focus just on the internal fluxes and examine the properties of

$$
\mathbf{S}_{\text{int}} : x_i \quad
\begin{array}{c}
v_i \\
\left(
\begin{array}{c}
 \\
 \\

\end{array}
\right)
\end{array}
$$

This form is useful to define pools of compounds that are conserved (see Chapter 10).

Example

The $\mathbf{S}_{\text{tot}}$ for the system shown earlier is given by

$$
\mathbf{S}_{\text{tot}} =
\begin{array}{cc}
\begin{array}{ccccc}
v_1 & v_2 & & b_1 & b_2
\end{array} & \\
\left(
\begin{array}{ccccc}
-1 & 1 & \vdots & 1 & 0 \\
1 & -1 & \vdots & 0 & -1 \\
\cdots & \cdots & \cdots & \cdots & \cdots \\
0 & 0 & \vdots & -1 & 0 \\
0 & 0 & \vdots & 0 & 1
\end{array}
\right) &
\begin{array}{c}
x_1 \\
x_2 \\
\\
c_1 \\
c_2
\end{array}
\end{array}
\tag{6.23}
$$

The internal stoichiometric matrix

$$\mathbf{S}_{\text{int}} = \begin{pmatrix} -1 & 1 \\ 1 & -1 \end{pmatrix}$$

has $m = 2$, $n = 2$, and $r = 1$. Thus all the fundamental subspaces have a dimension of 1.

The exchange stoichiometric matrix

$$\mathbf{S}_{\text{exch}} = \begin{pmatrix} -1 & 1 & \vdots & 1 & 0 \\ 1 & -1 & \vdots & 0 & -1 \end{pmatrix}$$

has $m = 2$, $n = 4$, and $r = 2$. It is full rank. The null space has a dimension of 2 ($= 4 - 2$), while the left null space has a dimension of 0 ($= 2 - 2$).

The total stoichiometric matrix has $m = 4$, $n = 4$, and $r = 3$. Thus, both of the null spaces are one-dimensional. In later chapters, we will learn how the dimensions of the null spaces relate to pools and pathways.

Partitioning $\mathbf{S}_{\text{int}}$ further

The internal stoichiometric matrix can be further partitioned. The compounds that cannot be exchanged with the environment form one group, and those that can, form another. In Chapter 10, for instance, we designate these two as *secondary* and *primary* metabolites, respectively.

Defining the system boundary

Note that in the earlier consideration we have drawn a systems boundary around the cell. Such definition is common since it is consistent with physical realities. However, since the definition of a systems boundary can be chosen, we can segment any network into subnetworks by drawing "virtual" boundaries. This property is useful in defining subsystems that may be "fast" (i.e., have rapid dynamics) and lead to temporal decomposition and subsystems that have a biochemical relevance (e.g., fatty acid biosynthesis).

6.7 Summary

➤ The stoichiometric matrix comprises stoichiometric coefficients that are commonly integer numbers.

➤ The columns of the stoichiometric matrix represent chemical reactions, while the rows represent compounds.

➤ The column vectors $\mathbf{s}_i$ represent chemical transformations and thus come with chemical information.

➤ The reaction vectors s_i imply elemental and charge balance. The reaction vectors are thus orthogonal to the rows of the elemental matrix. These conservation quantities are in the left null space of the stoichiometric matrix.

➤ Some quantities, such as free energy, are not conserved during a chemical reaction. These quantities will be in the row space of the stoichiometric matrix.

➤ There are few basic forms of the elementary reactions. Most, if not all, biochemical reaction networks are either linear or bilinear.

➤ Mathematically, the stoichiometric matrix is a linear mapping operation.

➤ Structurally, or topologically, the stoichiometric matrix represents a reaction map.

➤ The transpose of stoichiometric matrix represents a compound map.

➤ Both maps are topologically nonlinear, as they contain joint edges between nodes.

➤ The boundaries of a reaction network can be drawn in different ways and lead to three fundamental forms of **S**.

6.8   Further Reading

Arita M., "*In silico* atomic tracing by substrate–product relationships in *Escherichia coli* intermediary metabolism," *Genome Research*, **13**:2455–2466 (2003).

Lay, D.C., LINEAR ALGEBRA AND ITS APPLICATIONS, Addison-Wesley, New York (1994).

Schmidt, K., Carlsen, M., Nielsen, J., and Villadsen, J., "Modeling isotopomer distributions in biochemical networks using isotopomer mapping matrices," *Biotechnology & Bioengineering*, **55**:831–840, (1997).

Strang, G., LINEAR ALGEBRA AND ITS APPLICATIONS, 3rd edition, Harcourt, New York (1988).

CHAPTER 7

Topological Properties

The topological properties of matrices that describe the connectivity features of a network, such as the stoichiometric matrix, can be defined and studied. Elementary topological properties can be computed directly from the individual elements of **S**. Direct topological studies are interesting from a variety of standpoints. They focus on relatively easy to understand and intuitive properties of the structure of the network. Elementary topological properties relate to how connected a network is and how its components participate in forming the connectivity properties of the network. As pointed out in Chapter 2, there may be many functional states for a given network structure. Topological properties are thus global and less specific than functional states of networks. Some of the differences between functional states and network topology are covered in Part III.

7.1 The Binary Form of **S**

The elementary topological properties are determined based on the nonzero elements in the stoichiometric matrix. Thus, we define the elements of a new matrix $\hat{\mathbf{S}}$ as

$$\begin{aligned} \hat{s}_{ij} &= 0 \quad \text{if} \;\; s_{ij} = 0 \\ \hat{s}_{ij} &= 1 \quad \text{if} \;\; s_{ij} \neq 0 \end{aligned} \tag{7.1}$$

that is the *binary form* of **S**. This matrix comprises only 0's and 1's. If $\hat{s}_{ij}$ is unity, it means that compound i participates in reaction j. Note that in the rare case where a homodimer is formed, that is, in a reaction of the

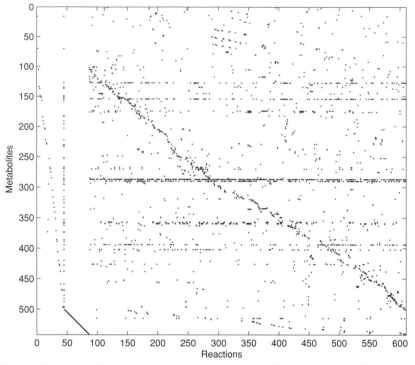

Figure 7.1: The stoichiometric matrix for *Geobacter sulfurreducens*. The dimensions of this matrix are $m = 541$ and $n = 609$, giving rise to 329,469 elements in the matrix. Of these, 2655, or 0.81%, of the elements are nonzero. Image provided by Radhakrishnan Mahadevan.

type $2A \rightarrow A_2$, the stoichiometric coefficient 2 becomes unity in the binary form of **S**.

S is a sparse matrix

A number of genome-scale stoichiometric matrices have been reconstructed (see Table 3.6). Since there are typically only two, three, or four compounds that participate in a reaction out of hundreds of compounds participating in a network, the stoichiometric matrix is *sparse*. A sparse matrix mostly comprises zero elements. For instance, if there are on average three compounds that participate in a reaction but there are m compounds in the network, then the fraction of nonzero elements in the matrix is $3/m$. If m is 300, then only 1% of the elements are nonzero and the matrix is sparse.

A pictorial representation of the genome-scale stoichiometric matrix for *Geobacter sulfurreducens* is shown in Figure 7.1. The 2,655 nonzero entries are indicated. They represent only 0.81% of the total of 329,469 elements in the matrix.

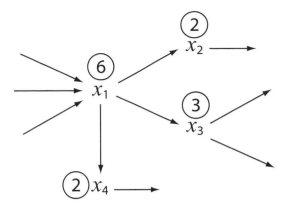

Figure 7.2: Simple reaction network illustrating connectivities of network nodes. The network shown has four nodes (compounds). The connectivities, ρ_i, for each node are given in the circled numbers next to a node. x_1 represents the most connected node.

7.2 Compound Participation and Connectivity

The number of nonzero entries in a row and a column of $\mathbf{S}$ give the two elementary topological properties. The sum of the nonzero entries in a column

$$\pi_j = \sum_{i=1}^{m} \hat{s}_{ij} \qquad (7.2)$$

gives the number of compounds that participate in reaction j. This quantity, π_j, can be called the *participation number* for a reaction. For elementary reactions, this number is most likely 3. Note that all compounds have to participate in a reaction for it to take place. It represents the number of nodes that form an edge in the reaction map.

The sum of the number of nonzero entries in a row

$$\rho_i = \sum_{j=1}^{n} \hat{s}_{ij} \qquad (7.3)$$

gives the number of reactions in which compound i participates. This number, ρ_i, is a measure of how connected, or linked, a compound is in the network. A compound that participates in a large number of reactions will form a highly connected node on the reaction map (see Figure 7.2). This number could be called the *connectivity number*, for the node, or simply its *connectivity*. Note that in a given functional state of a network, a compound does not have to participate in all the reactions that it is connected to, or to participate equally in them. This feature represents a fundamental difference between the flux and concentration maps.

Connectivities in genome-scale matrices

As soon as the first genome-scale matrices had been reconstructed (recall Table 3.6), the connectivities for all the metabolites were computed (see

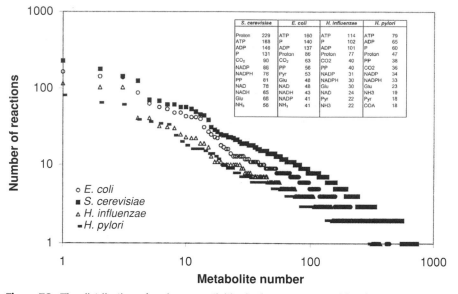

Figure 7.3: The distribution of node connectivities in the reconstructed first four genome-scale matrices. Data taken from [50, 51, 54, 67, 192].

Figure 7.3). Such computations show that there are relatively few metabolites (24 or so) that are highly connected, while most of the metabolites participate in only two reactions. This result is a reflection of the fact that few carrier molecules participate in a large number of reactions and a few metabolites are key to certain metabolic functions, such as nitrogen, one-carbon, and two-carbon metabolism (recall Table 3.1).

A surprising finding was the approximate linear appearance of the curve of the connectivities when the metabolites were rank ordered by decreasing connectivity, when plotted on a log–log scale [50] (see Figure 7.3). This curve can be redrawn based on the probability that a metabolite has a certain connectivity. Thus, there is a high probability of low connectivity and a low probability of high connectivity. Plotting these probabilities as a function of connectivity gives an approximate straight line on a log–log plot (Figure 7.4). Networks that show such a power law distribution are said to be *scale-free* [7].

The most highly connected nodes are carrier molecules, and due to their high connectivity they form the dominant features of **S** (see Chapter 8). Such biochemical insight has led to the analysis of the connectivity distributions of decomposed forms of **S** based on biochemical classification of reactions. Such analysis has concluded that genome-scale stoichiometric matrices are actually *scale-rich* and that the overall power law property is the result of the amalgamation of the connectivity properties of the biochemically classified modules [220]. Since functional states and their

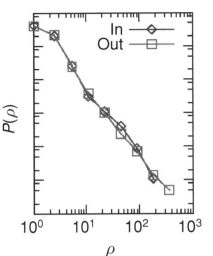

Figure 7.4: Connectivity probability distributions for metabolites. ρ represents the connectivity of a node, and $P(\rho)$ represents the probability that a node has connectivity ρ. The connectivity distribution is averaged over 43 different organisms. The connectivity of reactions into a node (In) and out of a node (Out) are represented separately. From [105].

biological interpretation is the focus of this text, the reader is directed to the list in Section 7.6 for more information on studies of network topological studies.

Biological interpretation

The biological significance of the power law distribution is not clear. It has been suggested that the most highly connected nodes in a network may represent the compounds that in the network were "first" in evolutionary time [239]. Such interpretations must of course be made in view of the constraining chemistry; for instance, there is a finite number of chemical transformations that a particular metabolite can undergo. Similarly, the "attack tolerance" of a network is such that the removal of the most highly connected nodes has the broadest impact on network functions [104, 105]. This consideration may apply to regulatory networks. Conversely, it is not possible to simply delete a metabolite from a network, but a link can be severed.

Node connectivity and network states

It should be noted that highly connected nodes may represent effective targets for drug development. However, topological properties of networks must be interpreted in the context of the more biologically relevant functional network states and their properties. One such consideration, for instance, is that a metabolic network must make all the biomass components of the cell in order for it to grow. Thus, even eliminating a step in a linear low-flux pathway leading to the synthesis of cofactors, vitamins, or amino acids will prevent a genome-scale metabolic network from supporting growth.

The relationship between node connectivity and lethality of reactions connecting into a node have been studied in genome-scale metabolic networks (Figure 7.5) [128]. The results in this figure show that the lethality fraction (f_L) of some of the less connected metabolites is higher than that of the highly connected metabolites irrespective of the size or the complexity of the metabolic network. In fact, surprisingly, for all of the networks studied, most of the points fall within a narrow range from 0.2 to 0.5. Links into poorly connected nodes are thus just as likely to be lethal as links into highly connected nodes. The number of lethal reactions around a highly connected metabolite such as OAA ($\rho_I = 10$) is shown in comparison to a poorly connected metabolite such as prbamp ($\rho_I = 2$) in Figure 7.5. However, the number of lethal reactions for both metabolites is 2, as prbamp occurs in a linear pathway in histidine biosynthesis and the deletion of either of the linked reactions leads to the loss of network function (growth). Therefore, although network topology as characterized by the elementary topological properties is certainly interesting and worthy of study, one must keep in mind the role of such properties with respect to attaining biologically meaningful functional states. The study of the functional states of networks is detailed in Part III of the text.

7.3   The Adjacency Matrices of S

An expanded set of elementary topological network properties can be obtained from the two adjacency matrices of $\hat{\mathbf{S}}$. One relates to the columns of $\hat{\mathbf{S}}$, while the other relates to the rows.

The reaction adjacency matrix $\mathbf{A}_v$

The premultiplication of a matrix by its transpose

$$\mathbf{A}_v = \hat{\mathbf{S}}^{\mathrm{T}}\hat{\mathbf{S}} \tag{7.4}$$

leads to a symmetrical matrix whose elements are the inner product of its columns, $\hat{\mathbf{s}}_i$. The diagonal elements of $\mathbf{A}_v$ are:

$$\langle \hat{\mathbf{s}}_i^{\mathrm{T}} \cdot \hat{\mathbf{s}}_i \rangle = \sum_k \hat{s}_{ki}^2 \tag{7.5}$$

Thus, since the elements of $\hat{\mathbf{s}}_i$ are 0 or 1, this summation simply represents the number of nonzero elements in $\hat{\mathbf{s}}_i$ or the number of compounds that participate in the reaction. The diagonal elements of $\mathbf{A}_v$ are thus the same quantity as given in equation 7.2.

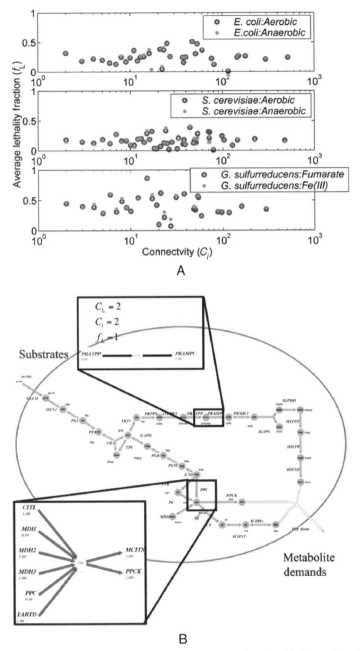

Figure 7.5: (A) The plot of the average lethality fraction ($f_{L,i}$) as a function of the metabolite connectivity (ρ_a) for the metabolic networks of *E. coli*, *S. cerevisiae*, and *G. sulfurreducens* under different growth conditions. (B) The reactions consuming or producing oxaloacetate (OAA, a key metabolite in the TCA cycle) and phosphoribosyl-AMP (prbamp, an intermediate in the histidine biosynthetic pathway) are shown. The reactions predicted to be essential for cell growth based on *in silico* analysis are shown as outlines, while the nonessential reactions are shown in black. The normalized predicted growth rate, connectivity (ρ_I), and the lethality fraction ($f_{L,i}$) are also shown. From [128].

The off-diagonal elements are given by

$$\langle \hat{\mathbf{s}}_j^T \cdot \hat{\mathbf{s}}_i \rangle = \sum_k \hat{s}_{jk} \hat{s}_{ki} \tag{7.6}$$

These elements can count how many compounds two reactions (reactions i and j) have in common.

The compound adjacency matrix, $\mathbf{A}_x$

The postmultiplication of a matrix by its transpose

$$\mathbf{A}_x = \hat{\mathbf{S}} \hat{\mathbf{S}}^T \tag{7.7}$$

leads to a symmetric matrix whose elements are the inner products of its rows. The diagonal elements are:

$$(\mathbf{a}_x)_{ii} = \sum_k \hat{s}_{ik}^2 \tag{7.8}$$

This summation gives the number of reactions in which compound x_i participates. This is the same quantity as computed in equation 7.3. The off-diagonal elements are

$$(\mathbf{a}_x)_{ij} = \sum_k s_{ik} s_{kj} \tag{7.9}$$

This is the number of reactions in which both compounds x_i and x_j participate and shows how extensively the two compounds are topologically connected in the network.

7.4 ████ Computation of the Adjacency Matrices

The reversible reaction

The stoichiometric matrix for a simple reversible reaction

$$\hat{\mathbf{S}} = \begin{pmatrix} 1 & 1 \\ 1 & 1 \end{pmatrix} \tag{7.10}$$

has two identical adjacency matrices:[1]

$$\mathbf{A}_v = \mathbf{A}_x = \begin{pmatrix} 2 & 2 \\ & 2 \end{pmatrix} \tag{7.11}$$

Thus, each compound participates in two reactions (the forward and backward reactions are treated separately), and there are two compounds participating in each reaction.

---

[1] Since the matrices $\mathbf{A}_v$ and $\mathbf{A}_x$ are symmetric, the elements below the diagonal are left blank.

The reversible bimolecular association reaction

The stoichiometric matrix for a reversible bimolecular association

$$\hat{\mathbf{S}} = \begin{pmatrix} 1 & 1 \\ 1 & 1 \\ 1 & 1 \end{pmatrix} \tag{7.12}$$

has adjacency matrices

$$\mathbf{A}_\nu = \begin{pmatrix} 3 & 3 \\ & 3 \end{pmatrix} \quad \text{and} \quad \mathbf{A}_x = \begin{pmatrix} 2 & 2 & 2 \\ & 2 & 2 \\ & & 2 \end{pmatrix} \tag{7.13}$$

Thus, $\mathbf{A}_\nu$ states that there are three compounds participating in each reaction (the forward and backward), and the two reactions have three compounds in common. Similarly, $\mathbf{A}_x$ states that each compound participates in two reactions (the diagonal) and that the first and second, first and third, and second and third compounds jointly participate in two reactions (forward and backward).

The reversible cofactor exchange reaction

The stoichiometric matrix for a reversible cofactor exchange reaction

$$\hat{\mathbf{S}} = \begin{pmatrix} 1 & 1 \\ 1 & 1 \\ 1 & 1 \\ 1 & 1 \end{pmatrix} \tag{7.14}$$

has adjacency matrices

$$\mathbf{A}_\nu = \begin{pmatrix} 4 & 4 \\ & 4 \end{pmatrix} \quad \text{and} \quad \mathbf{A}_x = \begin{pmatrix} 2 & 2 & 2 & 2 \\ & 2 & 2 & 2 \\ & & 2 & 2 \\ & & & 2 \end{pmatrix} \tag{7.15}$$

Thus, $\mathbf{A}_\nu$ states that there are four compounds participating in each reaction (forward and backward) and the two reactions have four compounds in common. Similarly, $\mathbf{A}_x$ states that each compound participates in two reactions (the diagonal) and that all pairwise combinations of the compounds jointly participate in two reactions (forward and backward).

Genome-scale matrices

The off-diagonal elements of $\mathbf{A}_x$ for the genome-scale metabolic networks for *Escherichia coli*, *Saccharonyces cerevisiae*, *Helicobacter pylori*, *Staphylococcus aureus*, and the human cardiac mitochondrion have been studied [11]. When rank ordered, they approximate a line on a log–log plot

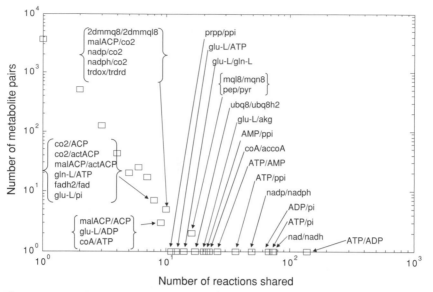

Figure 7.6: Metabolite coupling in *E. coli*. The number of metabolite pairs that share a given number of reactions are plotted and identified from the *E. coli* metabolic network. From [11].

for all networks considered. This suggests the notion of metabolite coupling, the concept that pairs of metabolites influence network behavior on different scales. The results for *E. coli* are shown in Figure 7.6. The pairs of metabolites that occur most often and dominate the network tend to be common cofactors (ATP/ADP, etc.). Other metabolite pairs have progressively less influence. The most often occurring metabolite pairs tend to be similar across the genome-scale networks studied, but their less prominent counterparts are often quite different.

7.5   Summary

➤ The binary form of **S** is $\hat{\mathbf{S}}$, which has zeros everywhere except unity, where a nonzero element appears in **S**.

➤ The summation of the elements in column j of $\hat{\mathbf{S}}$ give the number of compounds π_j that participate in reaction j.

➤ The summation of the elements in row i of $\hat{\mathbf{S}}$ give the number of reactions ρ_i in which compound i participates or shows how connected it is in the network.

➤ The binary stoichiometric matrix $\hat{\mathbf{S}}$ has two adjacency matrices $\mathbf{A}_v$ and $\mathbf{A}_x$ that are reaction and compound associated, respectively.

➤ A diagonal element of $\mathbf{A}_v$ gives the number of compounds (i.e., π_j) that participate in that reaction, and an off-diagonal element gives the number of compounds that the two corresponding reactions have in common.

➤ A diagonal element of $\mathbf{A}_x$ gives the number of reactions in which the corresponding compound participates (i.e., ρ_i), and an off-diagonal element gives the number of reactions in which the two corresponding compounds participate.

➤ The number of reactions that compounds participate in follow an approximate power law distribution in genome-scale matrices of metabolism. The number of reactions that pairs of metabolites participate in also follows a power law distribution.

7.6   Further Reading

Barabasi, A.L., LINKED: THE NEW SCIENCE OF NETWORKS, Perseus, Cambridge (2002).

Barabasi, A.L., and Oltavi, Z.N., "Network biology: Understanding the cell's functional organization," *Nature Reviews Genetics*, **5**:101–113 (2004).

Fell, D.A., and Wagner, A., "The small world of metabolism," *Nature Biotechnology*, **18**:1121-1122 (2000).

Mahadevan, K., and Palsson, B.O., "Properties of metabolic networks: Structure versus function," *Biophysical Journal*, **88**:L7–L9 (2005).

Tanaka, R., "Scale-rich metabolic networks," *Physical Review Letters*, **94**:168101–168104 (2005).

CHAPTER 8

Fundamental Subspaces of **S**

In the last chapter we discussed the elementary topological properties of the network that the stoichiometric matrix represents. In this chapter we look deeper into the properties of the stoichiometric matrix and how these fundamental topological properties can be used to obtain a more thorough understanding of the reaction network that it represents. This material is perhaps the most mathematical part of this book. It should be readily accessible to readers with formal education in the physical and engineering sciences, while readers with a life science background may find it challenging. The concepts introduced are important to the rest of the chapters in Part II. The stoichiometric matrix is a mathematical mapping operation (recall Figure 6.1). Matrices have certain fundamental properties that describe this mapping operation. These properties are contained in the four *fundamental subspaces* associated with a matrix. This chapter discusses these subspaces and how we can mathematically define them and interpret their contents in biochemical and biological terms.

8.1   Dimensions of the Fundamental Subspaces

The mapping that the stoichiometric matrix represents was illustrated in Figure 6.1 and a preliminary discussion of the associated four subspaces is found in Chapter 6. The stoichiometric matrix is typically rank deficient. The rank r of a matrix denotes the number of *linearly independent* rows and columns that the matrix contains. Rows are linearly dependent if any one row can be computed as a linear combination of the other rows. Linear dependency between the compounds and reactions determines the dimensionality of each of the four fundamental subspaces.

The dimensions of both the column and row space is r.

$$\dim(\text{Col}(\mathbf{S})) = \dim(\text{Row}(\mathbf{S})) = r$$

Since the dimension of the concentration vector is m, we have

$$\dim(\text{Left Null}(\mathbf{S})) = m - r$$

Similarly, the flux vector is n-dimensional; thus,

$$\dim(\text{Null}(\mathbf{S})) = n - r$$

The linear dependency between columns and rows of the stoichiometric matrix and its effect on the dimensionality of each fundamental subspace will be discussed further in subsequent chapters.

Contents of the fundamental subspaces

The four fundamental subspaces contain important information about a reaction network. Their contents are as follows:

- *Null space.* The null space of $\mathbf{S}$ contains all the steady-state flux distributions allowable in the network. The steady state is of much interest since most homeostatic states are close to being steady states.
- *Row space.* The row space of $\mathbf{S}$ contains all the dynamic flux distributions of a network and thus the thermodynamic driving forces that change the rate of reaction activity.
- *Left null space.* The left null space of $\mathbf{S}$ contains all the conservation relationships, or *time invariants*, that a network contains. The sum of conserved metabolites or conserved metabolic pools do not change with time and are combinations of concentration variables.
- *Column space.* The column space of $\mathbf{S}$ contains all the possible time derivatives of the concentration vector and thus shows how the thermodynamic driving forces move the concentration state of the network.

The contents of these spaces are described in detail in the subsequent chapters.

Basis for vector spaces

A *basis* for a space can be used to *span* the space. Thus, a basis describes all contents of a space. Different bases can be used for this purpose, including a linear basis, orthonormal basis (a special case of a linear basis) for linear spaces, and a convex basis for finite linear spaces. The choice of basis for the four fundamental subspaces becomes important since it influences the biological interpretation. Singular value decomposition gives simultaneous orthonormal bases for all the four fundamental subspaces.

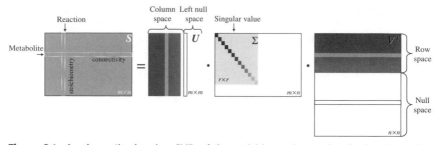

Figure 8.1: A schematic showing SVD of the stoichiometric matrix. The location of the orthonormal basis vectors for the four fundamental subspaces are indicated. Prepared by Iman Famili.

8.2   The Basics of Singular Value Decomposition

Singular value decomposition (SVD) is a well-established method used in a wide variety of applications, including signal processing, noise reduction, image processing, kinematics, and the analysis of high-throughput biological data [3, 91]. Unlike matrices that comprise experimentally determined numbers, the stoichiometric matrix is a "perfect" matrix that commonly comprises integers describing the structure of a reaction network. SVD of **S** can be used to analyze network properties, and it is a particularly useful way to obtain the basic information about the four fundamental subspaces of **S**.

SVD states that for a matrix **S** of dimension $m \times n$ and of rank r, there are orthonormal matrices **U** (of dimension $m \times m$) and **V** (of dimension $n \times n$) and a matrix with diagonal elements $\Sigma = \mathrm{diag}(\sigma_1, \sigma_2, \ldots, \sigma_r)$ with $\sigma_1 \geq \sigma_2 \geq \cdots \geq \sigma_r > 0$ such that

$$\mathbf{S} = \mathbf{U}\Sigma\mathbf{V}^{\mathrm{T}} \qquad (8.1)$$

SVD of **S** is shown schematically in Figure 8.1. The columns of **U** and **V** are the left and right singular vectors of **S**, respectively, and represent its *modes*, while the σ_i represent the singular values. The values in Σ give us the weight with which the modes contribute to the reconstruction of the matrix. These are rank ordered by decreasing magnitude in Σ with the largest singular value being first. For large systems, one can graph the magnitude of the r singular values to obtain a spectrum of singular values.

The singular value spectrum
The fractional singular values are calculated by

$$f_i = \frac{\sigma_i}{\sum_{k=1}^{r} \sigma_k} \qquad (8.2)$$

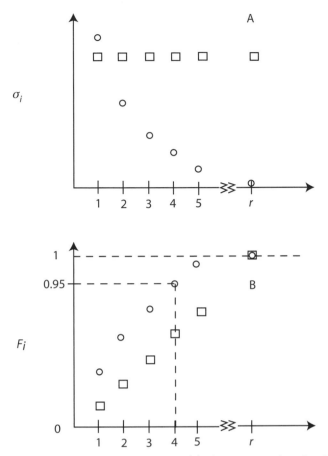

Figure 8.2: Singular value spectra. (A) The σ_i are rank ordered and plotted as a function of i. (B) Cumulative fractional singular values, F_i. One can use a certain fraction, i.e., 0.95, to determine the "effective dimensionality" representing 95% of the variance of the matrix being decomposed. Two types of spectra are illustrated. In the spectrum represented by the squares, the singular values are of similar magnitude, and thus the cumulative fractional singular values form a linear curve. Conversely, in the spectrum shown with the open circles, the relative magnitude of the singular values drops quickly, and the cumulative fractional spectrum rises quickly, leading to a low effective dimensionality of the mapping that the matrix represents.

Some example singular value spectra are given in Figure 8.2A. The cumulative fractional singular values, F_i, are defined as the sum of the first i fractional singular values:

$$F_i = \sum_{k=1}^{i} f_k \tag{8.3}$$

where i varies from 1 to r. In data analysis, one often uses a numerical criterion (i.e., 0.95) to terminate the cumulative spectrum and define the number of modes that generate 95% of the reconstruction of the

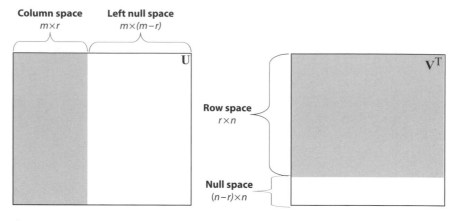

Figure 8.3: Orthonormal bases of the four fundamental subspaces of **S** obtained by SVD. Prepared by Iman Famili.

matrix. Since the stoichiometric matrix is of perfect precision, meaning no "measurement" noise in its elements, such cutoff may not be appropriate, depending on the information sought from the SVD. Some example cumulative spectra are given Figure 8.2B. Note that $F_r = 1$.

Orthonormal bases for the four fundamental subspaces

The columns of **U** are called the *left singular vectors* and the columns of **V** are the *right singular vectors*. The columns of **U** and **V** give orthonormal bases for all the four fundamental subspaces of **S** (see Figure 8.3). The first r columns of **U** and **V** give orthonormal bases for the column and row spaces, respectively. The last $m - r$ columns of **U** give an orthonormal basis for the left null space, and the last $n - r$ columns or **V** give an orthonormal basis for the null space.

The inner product of orthonormal vectors is zero. The inner product of an orthonormal vector with itself is unity. Thus,

$$\mathbf{U}^\mathrm{T}\mathbf{U} = \mathbf{I}_{(m \times m)} \quad \text{and} \quad \mathbf{V}^\mathrm{T}\mathbf{V} = \mathbf{I}_{(n \times n)} \tag{8.4}$$

The transposes of **U** and **V** are thus their inverses as well. The subscripts on the identity matrices in equation 8.4 are to remind the reader that they are not the same size.

Mapping between the singular vectors

The equation $\mathbf{S} = \mathbf{U}\mathbf{\Sigma}\mathbf{V}^\mathrm{T}$ can be rewritten as

$$\mathbf{SV} = \mathbf{U}\mathbf{\Sigma} \tag{8.5}$$

which can be expanded in terms of a series of independent equations that look like

$$\mathbf{S}\mathbf{v}_k = \sigma_k \mathbf{u}_k \tag{8.6}$$

In other words, $\mathbf{S}$ maps a right singular vector onto the corresponding left singular vector scaled by the corresponding singular value. A right singular vector (a column in $\mathbf{V}$) gives the weightings on the reaction vectors $\mathbf{s}_i$ needed to reconstruct each of the left singular vectors (a column in $\mathbf{U}$) as scaled by their respective singular values.

Mode-by-mode reconstruction of $\mathbf{S}$

The stoichiometric matrix can be reconstructed as

$$\mathbf{S} = \sum_{i=0}^{v} \sigma_i \left\langle \mathbf{u}_i \mathbf{v}_i^{\mathrm{T}} \right\rangle \tag{8.7}$$

where each term successively adds the contribution of each mode (or singular vector) to the reconstruction of $\mathbf{S}$.

A note on nomenclature

The naming conventions of the right singular vectors $\mathbf{v}_k$ and the flux vector $\mathbf{v}$ may cause confusion. Unfortunately, the literature uses the symbol $\mathbf{v}$ for both quantities, a convention that we will not change here. Both are vectors, denoted with a boldface font, but one has a subscript and the other does not. Equation 8.25 should help illustrate the difference between the two.

SVD as a series of transformations

Concentration variables can be transformed into groupings of concentrations that correspond to eigen-reactions, and these are driven by groupings of metabolic reactions. The basic mathematical nature of these transformations is shown in Figure 8.4. $\mathbf{V}^{\mathrm{T}}$ represents orthonormalization of the flux space, and these basis vectors are stretched by the singular values and mapped onto an orthonormal basis for the concentration space. The transformation $\mathbf{U}$ then converts the orthonormal concentrations back to the original coordinate system.

This set of transformations is conceptually useful. SVD has certain properties that make it convenient for numerical and mathematical analysis. The requirement for orthonormality makes chemical and biological interpretation difficult, and, as we will see in the subsequent chapters, an alternate set of basis vectors can be used to further such interpretations.

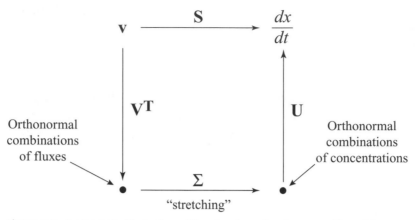

Figure 8.4: A schematic illustration of the singular value decomposition of **S**.

8.3 SVD of **S** for the Elementary Reactions

To develop an understanding of the information that SVD of **S** provides, we will apply it to the elementary reactions introduced in Chapter 6. This will show that the fundamental subspaces are finite and for some purposes may better be spanned by a nonorthonormal set of basis vectors.

Reversible conversion

Consider the reaction of equation 6.11 written as

$$x_1 \underset{v_2}{\overset{v_1}{\rightleftharpoons}} x_2 \tag{8.8}$$

where $x_1 = CP$ and $x_2 = PC$. The corresponding stoichiometric matrix can be decomposed as

$$\mathbf{S} = \mathbf{U\Sigma V}^{\mathrm{T}}$$

or

$$\begin{pmatrix} -1 & 1 \\ 1 & -1 \end{pmatrix} = \frac{1}{\sqrt{2}} \begin{pmatrix} -1 & 1 \\ 1 & 1 \end{pmatrix} \begin{pmatrix} 2 & 0 \\ 0 & 0 \end{pmatrix} \frac{1}{\sqrt{2}} \begin{pmatrix} 1 & -1 \\ 1 & 1 \end{pmatrix} \tag{8.9}$$

The four fundamental subspaces are shown in Figure 8.5. They are all one dimensional. Thus, the second column of **U** spans the left null space and, as we will see in Chapter 10, corresponds to a conservation relationship, $x_1 + x_2$ being a constant. Similarly, the second row of $\mathbf{V}^{\mathrm{T}}$ spans the null space of the stoichiometric matrix, and, as we will see in Chapter 9, it corresponds to a type III circular extreme pathway.

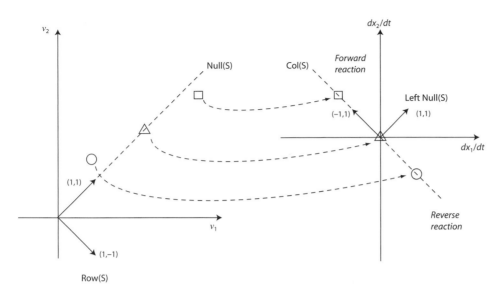

Figure 8.5: The four fundamental subspaces of the stoichiometric matrix for the reaction $x_1 \rightleftharpoons x_2$. Here, $n = m = 2$ and $r = 1$. Thus all four fundamental subspaces are one dimensional. Note that we have multiplied all the basis vectors by $\sqrt{2}$ to make it easier to visualize the direction of the basis vectors of the four subspaces.

The column and row spaces are one dimensional and related by

$$\mathbf{S}\mathbf{v}_1 = \sigma_1\mathbf{u}_1$$

or

$$\mathbf{S}\begin{pmatrix} 1/\sqrt{2} \\ -1/\sqrt{2} \end{pmatrix} = 2\begin{pmatrix} -1/\sqrt{2} \\ 1/\sqrt{2} \end{pmatrix} \quad \text{or} \quad \mathbf{S}\begin{pmatrix} 1 \\ -1 \end{pmatrix} = 2\begin{pmatrix} -1 \\ 1 \end{pmatrix} \tag{8.10}$$

The row space is spanned by $\mathbf{v}_1^T = (1, -1)/\sqrt{2}$, meaning that for a net flux through the reaction, the time derivatives in the column space are moved in the opposite direction multiplied by a factor of 2. As shown in Figure 8.5, if (v_1, v_2) is located above the 45-degree line, the distance from the 45-degree line is doubled and projected in the opposite direction in the time derivative space. The opposite is true for a point located below the 45-degree line. If the numerical values of v_1 and v_2 are the same, there is no net reaction and the time derivatives are zero.

The finite size of the fundamental subspaces

All fluxes of elementary reactions are positive and have a maximal rate. Thus, all flux values fall into a range $0 \leq v_i \leq v_{i,\max}$. This range for v_1 and v_2 for the elementary reaction $x_1 \rightleftharpoons x_2$ is shown in Figure 8.6. The null space is on a diagonal line, while the rest of the square is the possible row space. Every point in the square can be decomposed into a steady state (v_{ss}) and a dynamic (v_{dyn}) component. These two components are represented

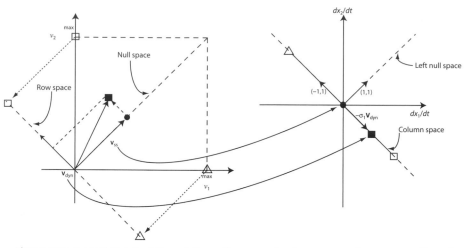

Figure 8.6: A graphical depiction of the null, row, and column spaces for $x_1 \rightleftharpoons x_2$. Since the fluxes v_1 and v_2 are finite, all these three spaces are finite. The singular vectors shown are multiplied by $\sqrt{2}$ to make the figure more simplistic.

by the $(1, 1)/\sqrt{2}$ and $(1,-1)/\sqrt{2}$ vectors. They are orthogonal and span the null and row spaces respectively; v_{ss} is mapped into the origin by **S**, whereas v_{dyn} is mapped onto $\mathbf{u}_1 = (-1, 1)/\sqrt{2}$ and stretched by the singular value (see equation 8.10).

Note that the bounded range of the fluxes also set the bounds of the column space. The extreme points of the row space (the open triangle and square) correspond to the maximum allowable values on the time derivatives of x_1 and x_2. Thus, the extreme points of the row space lead to extreme points in the column space. The implication of nonnegativity and finite size of the flux and concentration values will be discussed in subsequent chapters.

Numerical example

We can trace these mappings using a specific numerical example. If we pick $\mathbf{v} = (2\sqrt{2}, \sqrt{2})^{\mathrm{T}}$, then

$$\mathbf{V}^{\mathrm{T}}\mathbf{v} = \begin{pmatrix} 1 \\ 3 \end{pmatrix} \tag{8.11}$$

which corresponds to the projection of **v** onto the two right singular vectors. In other words, this flux vector is decomposed into one unit of v_1 and three of v_2. Then

$$\Sigma\mathbf{V}^{\mathrm{T}}\mathbf{v} = \begin{pmatrix} 2 \\ 0 \end{pmatrix} \tag{8.12}$$

which finally maps onto the left singular vectors as

$$\dot{x} = \mathbf{U}\Sigma\mathbf{V}^{\mathsf{T}}\mathbf{v} = \mathbf{U}\begin{pmatrix} 2 \\ 0 \end{pmatrix} = 2 \cdot \mathbf{u}_1 + 0 \cdot \mathbf{u}_2 = 2\mathbf{u}_1 = \begin{pmatrix} -\sqrt{2} \\ \sqrt{2} \end{pmatrix} \qquad (8.13)$$

Bilinear association

Consider the reaction of equation 6.13 written as

$$x_1 + x_2 \underset{v_2}{\overset{v_1}{\rightleftharpoons}} x_3 \qquad (8.14)$$

where $x_1 = C$, $x_2 = P$ and $x_3 = CP$. The SVD of the stoichiometric matrix is

$$\begin{pmatrix} -1 & 1 \\ -1 & 1 \\ 1 & -1 \end{pmatrix} = \begin{pmatrix} -1/\sqrt{3} & 2/\sqrt{6} & 0 \\ -1/\sqrt{3} & -1/\sqrt{6} & 1/\sqrt{2} \\ 1/\sqrt{3} & 1/\sqrt{6} & 1/\sqrt{2} \end{pmatrix} \begin{pmatrix} \sqrt{6} & 0 \\ 0 & 0 \\ 0 & 0 \end{pmatrix} \begin{pmatrix} 1/\sqrt{2} & -1/\sqrt{2} \\ -1/\sqrt{2} & -1/\sqrt{2} \end{pmatrix}$$

There is only one singular value, and thus the column space is spanned by one left singular vector $\mathbf{u}_1$. It is the reaction vector $\mathbf{s}_i$ normalized to unit length. The row space is spanned by $\mathbf{v}_1^{\mathsf{T}} = (1, -1)/\sqrt{2}$. The row and column spaces are related by

$$\mathbf{S}\mathbf{v}_1 = \sigma_1 \mathbf{u}_1 \qquad (8.15)$$

or

$$\begin{pmatrix} -1 & 1 \\ -1 & 1 \\ 1 & -1 \end{pmatrix} \begin{pmatrix} 1/\sqrt{2} \\ -1/\sqrt{2} \end{pmatrix} = \sqrt{6} \begin{pmatrix} -1/\sqrt{3} \\ -1/\sqrt{3} \\ 1/\sqrt{3} \end{pmatrix} \qquad (8.16)$$

or

$$\begin{pmatrix} -1 & 1 \\ -1 & 1 \\ 1 & -1 \end{pmatrix} \begin{pmatrix} 1 \\ -1 \end{pmatrix} = 2 \begin{pmatrix} -1 \\ -1 \\ 1 \end{pmatrix} \qquad (8.17)$$

Note that the row and null spaces are spanned by the same right singular vectors, $\mathbf{v}_1^{\mathsf{T}}$ and $\mathbf{v}_2^{\mathsf{T}}$, respectively, as the reversible conversion. This same decomposition is true of the flux vector, leading to an analogous interpretation. The column space is simply spanned by the normalized form of the reaction vector. The left null space is now two dimensional. The orthonormal basis vectors for the column and left null space are shown in Figure 8.7. The second and third left singular vectors $\mathbf{u}_2$ and $\mathbf{u}_3$ that span the left null space are hard to interpret chemically. We will address this issue in what follows.

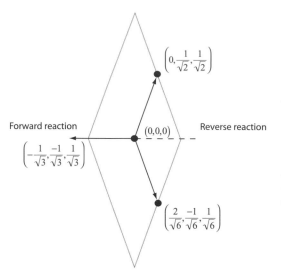

Figure 8.7: The three-dimensional depiction of the orthonormal basis for the column and left null spaces for a simple bilinear association. The plane is the left null space and the line is the column space. The vectors shown are an orthonormal set. If the flux vector is on the left-hand side of the plane as indicated, then the reaction is proceeding in the forward direction, and vice versa.

Linear combinations of fluxes and concentrations

We can begin to familiarize ourselves with the details of these transformations. The flux vector in the dynamic equations can be transformed using $\mathbf{V}^\mathrm{T}$ as

$$\frac{d\mathbf{x}}{dt} = \mathbf{SVV}^\mathrm{T}\mathbf{v} \tag{8.18}$$

or

$$\frac{d}{dt}\begin{pmatrix} x_1 \\ x_2 \\ x_3 \end{pmatrix} = \begin{pmatrix} -1 \\ -1 \\ 1 \end{pmatrix}(v_1 - v_2) - \begin{pmatrix} 0 \\ 0 \\ 0 \end{pmatrix}(v_1 + v_2) \tag{8.19}$$

forming two groupings of the fluxes. The second term corresponds to the null space, and the combination $v_1 + v_2$ is a type III extreme pathway that we will discuss in Chapter 9. The first corresponds to the row space and the grouping $v_1 - v_2$ is the net flux through the reaction, and it is orthogonal to $v_1 + v_2$ (see Figure 8.5). Multiplying equation 8.19 by $\mathbf{U}^\mathrm{T}$ leads to

$$\frac{d}{dt}\begin{pmatrix} (-x_1 - x_2 + x_3)/\sqrt{3} \\ (2x_1 - x_2 + x_3)/\sqrt{6} \\ (x_2 + x_3)/\sqrt{2} \end{pmatrix} = \begin{pmatrix} \sqrt{6} \\ 0 \\ 0 \end{pmatrix}\frac{(v_1 - v_2)}{\sqrt{2}} - \begin{pmatrix} 0 \\ 0 \\ 0 \end{pmatrix}\frac{(v_1 + v_2)}{\sqrt{2}}$$

Note that the singular value of $\sqrt{6}$ shows up and that the two column vectors on the right-hand side of the equation are the two columns of Σ.

This system is now fully decomposed, showing how independent groupings of concentrations are moved by independent groupings of the fluxes. As noted earlier, the two left singular vectors that span the left null

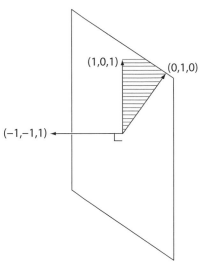

Figure 8.8: The three-dimensional depiction of a convex basis for the column space and chemical moiety-based basis for the left null space for a simple bilinear association. Note that the two basis vectors ((1,0,1) and (0,1,1)) for the left null space are not orthogonal to each other, but both are orthogonal to the basis vector for the column space. The wedge shown contains the nonnegative combinations of the two basis vectors.

space are not easy to interpret chemically. Since they are not changed by the groupings of fluxes (0's in the last two rows of the vectors on the right-hand side), they can be combined without changing the dynamic solution.

Nonorthonormal basis vectors

The second and third left singular vectors can be combined to give $(1, 0, 1)$ and $(0, 1, 1)$ that still span the left null space. These are not orthonormal vectors spanning the left null space, but they represent chemical conservation moieties, or pools, which are $x_1 + x_3$ (the moiety C based on equation 6.13) and $x_2 + x_3$ (the moiety P based on equation 6.13), respectively. This basis for the left null space is shown in Figure 8.8. Note that the segment of the left null space that is chemically meaningful lies in the wedge spanned by these two vectors since only a nonnegative combination of them is chemically possible. This basis is called a *convex basis* for the left null space, and we will discuss these issues in more detail in Chapter 10. Thus, although mathematically convenient and useful, the use of the orthonormal bases obtained by SVD may not be well suited for chemical and biological interpretation of the left null space. We will also see in Chapter 9 that convex bases for the null space are quite useful.

8.4   Interpretation of SVD: Systemic Reactions

The orthonormal basis vectors that are obtained from SVD do give useful information about the properties of the overall chemical transformations that characterize a network. The basic dynamic mass balance equation

$$\frac{d\mathbf{x}}{dt} = \mathbf{Sv} \tag{8.20}$$

can be rearranged as

$$\mathbf{U}^T \frac{d\mathbf{x}}{dt} = \mathbf{U}^T \mathbf{SVV}^T \mathbf{v} \tag{8.21}$$

or

$$\frac{d(\mathbf{U}^T \mathbf{x})}{dt} = \Sigma(\mathbf{V}^T \mathbf{v}) \tag{8.22}$$

Thus, the left singular vectors ($\mathbf{u}_i$) form linear combinations of the concentration variables and the right singular vectors ($\mathbf{v}_i$) form linear combinations of the fluxes. Groups of concentration variables are called *pools*, and groups of fluxes are called *pathways* (see Chapters 9 and 10, respectively).

The dynamic relationship between the groupings of fluxes and concentrations that correspond to the nonzero singular values can be written as

$$\frac{d(\mathbf{u}_k^T \mathbf{x})}{dt} = \sigma_k(\mathbf{v}_k^T \mathbf{v}) \tag{8.23}$$

This simple derivation shows that there is a linear combination of compounds:

$$\mathbf{u}_k^T \mathbf{x} = u_{k1}x_1 + u_{k2}x_2 + \cdots + u_{kr}x_m \tag{8.24}$$

which is being uniquely moved by a linear combination of metabolic fluxes as

$$\mathbf{v}_k^T \mathbf{v} = v_{k1}v_1 + v_{k2}v_2 + \cdots + v_{kn}v_n \tag{8.25}$$

and the extent of this motion is given by σ_k. An important feature of SVD is that the singular vectors are orthonormal to each other, and consequently each of the kth motions in equations 8.24 and 8.25 are structurally decoupled.

Equation 8.23, therefore, defines an *eigen-reaction* or a *systemic metabolic reaction* as

$$\Sigma u_{ki}x_i \underset{\substack{\Sigma v_{kj}v_j \\ \text{for } v_{kj} < 0}}{\overset{\substack{\Sigma v_{kj}v_j \\ \text{for } v_{kj} > 0}}{\rightleftharpoons}} \Sigma u_{ki}x_i \tag{8.26}$$

where the elements of $\mathbf{u}_k$ are equivalent to *systemic stoichiometric coefficients* and the elements of $\mathbf{v}_k$ are *systemic participation numbers*. Note that the $\mathbf{u}_k$ vectors correspond to systemic reaction vectors that are analogous to

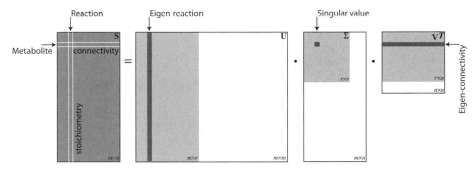

Figure 8.9: A schematic of the singular value decomposition of the stoichiometric matrix. Prepared by Iman Famili.

$\mathbf{s}_i$ (see Figure 8.9). Thus, as we move a point along this vector, compounds with negative u_{ki} values decrease, while those with positive u_{ki} increase, and vice versa. Similarly, the reactions with positive v_{kj} values will drive a point in the increasing direction of $\mathbf{u}_k$, while those with negative values will act in the opposite direction. This relationship is graphically illustrated in Figure 8.10. Thus, equation 8.26 describes a systemic reaction. These systemic metabolic reactions can be used to describe the function of the network as a whole.

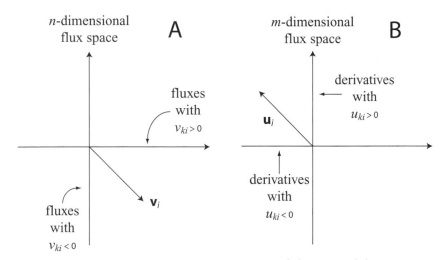

Figure 8.10: The relationship between corresponding left ($\mathbf{u}_i$) and right ($\mathbf{v}_i$) singular vectors as a systemic reaction. The right singular vector can be broken up into two parts, containing positive elements (on x-axis) and negative elements (on y-axis) (A). Reactions with positive elements correspond to reactions driving the systemic reaction forward, while those with negative elements drive it in the reverse direction. Analogously, the left singular vector $\mathbf{u}_i$ can be broken into a part with positive elements (y-axis) and negative elements (x-axis) (B). The former corresponds to compounds formed by the systemic reaction, while the latter corresponds to those disappearing. Since $\mathbf{S}$ maps $\mathbf{v}_i$ onto $\mathbf{u}_i$, all points on $\mathbf{v}_i$ correspond to a point on $\mathbf{u}_i$. Further, since $\mathbf{v}_i$ and $\mathbf{v}_j$ are orthonormal, the systemic reactions are independent.

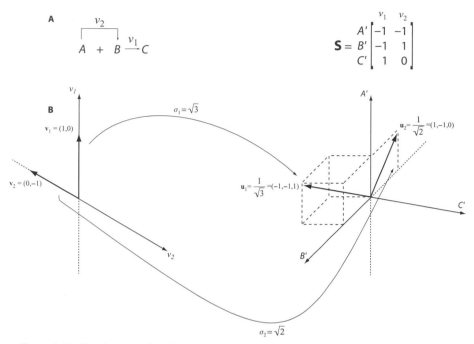

Figure 8.11: Simple example of SVD analysis for reacting systems. (A) Schematic of a reaction network with two reactions (v_1 and v_2) that act on metabolites A, B, and C, and its corresponding stoichiometric matrix, **S**. (B) Graphical representation of the mapping between the eigen-connectivities and eigen-reactions. The two singular vectors that correspond to v_1 and v_2 are orthogonal and catalyze each eigen-reaction, $\mathbf{u}_1$ and $\mathbf{u}_2$, independently. The two singular vectors of time derivatives are orthogonal in a three-dimensional space. The magnitude of the singular values σ_1 and σ_2 indicates the relative contribution of its corresponding singular vectors to the overall construct of the biochemical transformation in the network (A', time derivative of A). This reaction schema is similar to the reactions in glycolysis where a hexose (FDP) is split into two trioses (GAP and DHAP), which are interconverted by an isomerase. From [57].

Simple example

These general concepts can be illustrated by a simple example (Figure 8.11). The two reactions, v_1 and v_2, in Figure 8.11A form a two-dimensional flux space and relate metabolites A, B, and C in the space where the time derivatives of concentrations lie. The two singular vectors $\mathbf{v}_1$ and $\mathbf{v}_2$ are orthogonal and drive the singular vectors of time derivatives, $\mathbf{u}_1$ and $\mathbf{u}_2$, independently (Figure 8.11B). Note that reaction v_2 only drives the motion of A and B. The singular values in this simple example are $\sigma_1 = 1.73$ and $\sigma_2 = 1.41$ and indicate the relative contribution of each singular vector to the overall construct of the biochemical motion in the network.

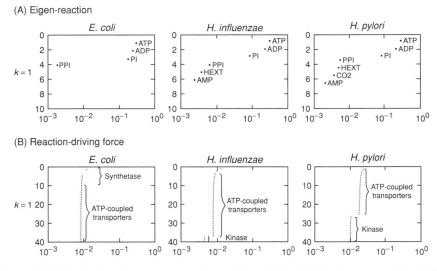

Figure 8.12: The first mode in genome-scale matrices of *H. pylori*, *H. influenzae*, and *E. coli*. (A) The compound composition of the mode; (B) the reactions driving the conversion. Note that the values on the x-axis are shown by u_{ij}^2 and v_{ij}^2 to make them all positive. From [57].

Decomposition of genome-scale matrices

The genome-scale matrices for *Helicobacter pylori*, *Haemophilus influenzae*, and *Escherichia coli* (see Table 3.6) have been studied using SVD [57]. The dominant four modes of the three matrices were similar, accounting for about 27% of the cumulative singular value spectra. The cofactors participating in energy, redox, and phosphate metabolism emerge with the most significant values in the first four eigen-reactions of all three genome-scale networks.

- The first eigen-reaction in all the genomes is the conversion of ATP to ADP and P_i (see Figure 8.12).
- The second eigen-reaction describes the conversion of NADP to NADPH (see Figure 8.13).

Although the eigen-reactions are similar in the three organisms, the metabolic reactions participating in driving the eigen-reactions differ somewhat from one network to another. This difference is in part due to dissimilarity among metabolic reactions in these organisms. The reaction participation in the two principal conversions is as follows:

- The ATP-coupled transporters have the highest participation numbers of the first right singular vectors, except for *E. coli* where a group of synthetase reactions is present. A group of kinases with identical

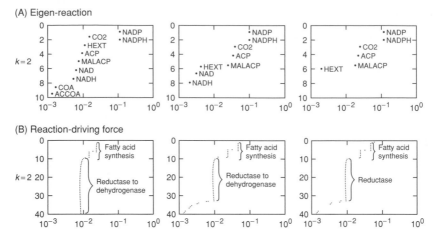

Figure 8.13: The second mode in genome-scale matrices of *H. pylori, H. influenzae,* and *E. coli.* (A) The reaction composition of the mode; (B) the reactions driving the conversion. Note that the values on the x-axis are shown by u_{ij}^2 and v_{ij}^2 to make them all positive. From [57].

participation numbers follow. These three types of reactions thus define the dominant eigen-reaction in all three genomes.

- The second mode corresponds to redox conversions that involve NADPH. In all three genomes, fatty acid synthesis reactions have dominant participation numbers in this mode. Then, a group of reductases appears with 5- to 10-fold lower participation numbers. These two types of redox exchanges dominate NADPH metabolism.

Thus, the dominant eigen-reactions correspond to important generic biochemical transformations of metabolites in all these organisms, but the sets of metabolic reactions that participate in these transformations may differ significantly in a number of modes while being similar in others. The overall dominant features of these three genome-scale networks thus represent similar but not identical chemical transformations.

8.5   Summary

➤ SVD provides unbiased and decoupled information about all the fundamental subspaces of **S** simultaneously.

➤ The first r columns of the left singular matrix **U** contain a basis for the column space of **S**, and the remaining $m - r$ columns contain a basis for the left null space.

➤ The first r columns of the right singular matrix **V** contain a basis for the row space of **S** and the remaining $n - r$ columns contain a basis for the null space.

➤ The sets of basis vectors in **U** and **V** are orthonormal.

➤ The first r columns of **U** give systemic reactions, analogous to a single column of **S**, representing a single reaction.

➤ The corresponding column of **V** gives the combination of the reactions that drive a systemic reaction.

➤ Orthonormal basis vectors are mathematically convenient but not necessarily biologically or chemically meaningful.

8.6 Further Reading

Famili, I., and Palsson B.O., "Systemic metabolic reactions are obtained by singular value decomposition of genome-scale stoichiometric matrices," *Journal of Theoretical Biology*, **224**:87–96 (2003).

Meyer, C.D., MATRIX ANALYSIS AND APPLIED LINEAR ALGEBRA, SIAM, Philadelphia (2000).

CHAPTER 9

The (Right) Null Space of **S**

The right null space of the stoichiometric matrix contains the steady-state flux distributions through the network that the matrix represents. It is typically just called the *null space*. The choice of basis for the null space is important in describing its contents in meaningful chemical and biological terms. A convex representation of the null space has proven useful to meet this goal. The convex basis vectors correspond to pathways through a reconstructed network. We will introduce the basic concepts in this chapter and describe their use for large-scale networks in Chapter 13.

9.1 Definition

The right null space of **S** is defined by

$$\mathbf{S}\mathbf{v}_{ss} = \mathbf{0} \tag{9.1}$$

Thus, all the steady-state flux distributions, $\mathbf{v}_{ss}$, are found in the null space. The null space has a dimension of $n - r$. Note that $\mathbf{v}_{ss}$ must be orthogonal to all the rows of **S** simultaneously and thus represents a linear combination of flux values on the reaction map that sum to zero. Therefore, the null space is orthogonal to the row space of **S** (recall Figure 6.1).

The right null space is spanned by a set of $n - r$ basis vectors, $\mathbf{r}_i, i \in [1, n - r]$. The set of basis vectors form columns of a matrix **R** that satisfy

$$\mathbf{S}\mathbf{R} = \mathbf{0} \tag{9.2}$$

that can be pictorially (where horizontal lines are row vectors, and vertical lines are column vectors) represented as

$$(\equiv)(|||) = \mathbf{0} \quad \text{with} \quad \mathbf{R} = (|||) \quad \text{thus any} \quad \mathbf{v}_{ss} = \Sigma w_i \mathbf{r}_i \tag{9.3}$$

A set of linear basis vectors is not unique, but once the set is chosen, the weights (w_i) for a particular $\mathbf{v}_{ss}$ are unique.

A basis vector $\mathbf{r}_i$ makes nodes in the flux map *link neutral* (i.e., balanced) since it is orthogonal to all the rows of $\mathbf{S}$ simultaneously. Such network-scale flux balancing may be nonintuitive for topologically nonlinear networks. This balancing requirement leads to *network-based* pathway definitions and the use of basis vectors ($\mathbf{p}_i$) that represent such pathways.

If a reacting system is closed, only internal cycles are possible. These cycles are at a thermodynamic equilibrium and thus of little biological interest. Living systems are open to the environment, and thus it is the trafficking of substances in and out of a cell that becomes a key interest. This feature leads to the segregation of the flux vector into two parts, internal and external fluxes, as discussed in Chapter 6. The internal fluxes can be defined by elementary reactions that have nonnegative flux values and thus are naturally considered to be unidirectional. The exchange fluxes can involve diffusive processes and may not have a natural representation as chemical reactions. The exchange fluxes are thus naturally considered to be bidirectional.

9.2   Choice of Basis

A linear basis for the null space can be computed using a number of standard methods, including SVD (see Chapter 8). An infinite number of different bases exist for a linear space. We are interested in finding a basis that is biochemically meaningful and thus useful for biological interpretation.

Linear basis

A simple linear reaction network is shown in Figure 9.1A. The null space is defined by

$$
\begin{pmatrix}
1 & -1 & 0 & 0 & -1 & 0 \\
0 & 1 & -1 & 0 & 0 & 0 \\
0 & 0 & 1 & -1 & 0 & 1 \\
0 & 0 & 0 & 0 & 1 & -1
\end{pmatrix}
\begin{pmatrix}
v_1 \\ v_2 \\ v_3 \\ v_4 \\ v_5 \\ v_6
\end{pmatrix}
=
\begin{pmatrix}
0 \\ 0 \\ 0 \\ 0
\end{pmatrix}
\tag{9.4}
$$

The matrix is full rank, and thus the dimension of the null space is $2\,(= 6 - 4)$. Since columns 4 and 6 do not contain pivots, this set of linear

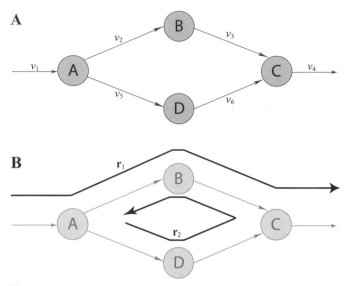

Figure 9.1: Simple network and basis vectors for the null space of the corresponding **S**. (A) The reaction map; (B) the basis vectors from equation 9.5 drawn as flux distributions on the reaction map. Prepared by Christophe Schilling.

equations can be solved using v_4 and v_6 as the free variables to give

$$
\begin{pmatrix} v_1 \\ v_2 \\ v_3 \\ v_4 \\ v_5 \\ v_6 \end{pmatrix} = \begin{pmatrix} v_4 \\ v_4 - v_6 \\ v_4 - v_6 \\ v_4 \\ v_4 \\ v_6 \\ v_6 \end{pmatrix} = v_4 \begin{pmatrix} 1 \\ 1 \\ 1 \\ 1 \\ 0 \\ 0 \end{pmatrix} + v_6 \begin{pmatrix} 0 \\ -1 \\ -1 \\ 0 \\ 1 \\ 1 \end{pmatrix} = w_1 \mathbf{r}_1 + w_2 \mathbf{r}_2 \qquad (9.5)
$$

where $\mathbf{r}_1$ and $\mathbf{r}_2$ form a basis. They are both in the null space and are linearly independent. For any numerical values of v_4 and v_6, a flux vector will be computed that lies in the null space. These basis vectors can be shown schematically by graphing them on to the reaction map (Figure 9.1B).

Any steady-state flux distribution is a unique linear combination of the two basis vectors. For example,

$$
\mathbf{v} = \begin{pmatrix} 2 \\ 1 \\ 1 \\ 2 \\ 1 \\ 1 \end{pmatrix} = w_1 \mathbf{r}_1 + w_2 \mathbf{r}_2 = (2) \begin{pmatrix} 1 \\ 1 \\ 1 \\ 1 \\ 0 \\ 0 \end{pmatrix} + (1) \begin{pmatrix} 0 \\ -1 \\ -1 \\ 0 \\ 1 \\ 1 \end{pmatrix} = (2)\mathbf{r}_1 + (1)\mathbf{r}_2 \qquad (9.6)
$$

This combination can also be drawn on the reaction map (Figure 9.2A). This set of basis vectors, although mathematically valid, is chemically

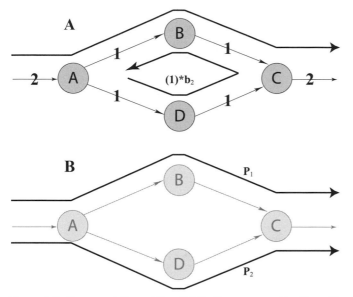

Figure 9.2: Representation of flux distribution as a combination of basis vectors (from equation 9.5 in (A); and of a set of nonnegative basis vectors (from equation 9.7, in (B). Prepared by Christophe Schilling.

unsatisfactory. The reason is that the second basis vector, r_2, represents fluxes through irreversible elementary reactions, v_2 and v_3, in the reverse direction, and it thus represents a chemically unrealistic event.

Nonnegative linear basis

The problem with the acceptability of the basis just considered stems from the fact that the flux through an elementary reaction can only be positive, i.e., $v_i \geq 0$. A negative coefficient in the corresponding row in the basis vector that multiplies the flux is thus undesirable. This consideration leads to the need to have nonnegative basis vectors for the null space. In the earlier example we can combine the basis vectors to eliminate all negative elements in them. This combination is achieved by transforming the set of basis vectors by

$$
(\mathbf{r}_1, \mathbf{r}_2) =
\begin{pmatrix}
1 & 0 \\
1 & -1 \\
1 & -1 \\
1 & 0 \\
0 & 1 \\
0 & 1
\end{pmatrix}
\begin{pmatrix}
1 & 1 \\
0 & 1
\end{pmatrix}
=
\begin{pmatrix}
1 & 1 \\
1 & 0 \\
1 & 0 \\
1 & 1 \\
0 & 1 \\
0 & 1
\end{pmatrix}
= (\mathbf{p}_1, \mathbf{p}_2) \qquad (9.7)
$$

In this new basis, $\mathbf{p}_1$, $\mathbf{p}_2$, the first basis vector is the same as in the old basis, whereas the second basis vector in the new set is an addition of the two basis vectors in the old basis. These two new basis vectors are shown on

Table 9.1: Comparing the properties of linear and convex bases.

Linear spaces	Convex spaces
Described by linear equations	Described by linear equations and inequalities
Vector spaces defined by a set of linearly independent basis vectors ($\mathbf{b}_i$)	Convex polyhedral cone defined by a set of conically independent generating vectors ($\mathbf{p}_i$)
$\mathbf{v} = \sum w_i \mathbf{b}_i \quad -\infty \leq w_i \leq +\infty$	$\mathbf{v} = \sum \alpha_i \mathbf{p}_i \quad 0 \leq \alpha_i \leq +\infty$
Every point in the vector space is uniquely described by a linear combination of basis vectors (unique representation for a given basis)	Every point in the vector space is described as a nonnegative linear combination of the generating vectors (nonunique representation)
Number of basis vectors equals dimension of the null space	Number of generating vectors may exceed dimension of the null space
Infinite number of bases that can be used to span the space	Unique set of generating vectors

the reaction map in Figure 9.2B, and they contain no fluxes that operate in the incorrect direction. Notice that these nonnegative basis vectors look like pathways through this simple system. We remind the reader that this toy system is biochemically irrelevant since there are no carrier or cofactor exchange reactions.

Convex versus linear bases

The introduction of nonnegative basis vectors leads us to convex analysis. Convex analysis is based on equalities (in our case, $\mathbf{Sv} = \mathbf{0}$) and inequalities (in our case, $0 \leq v_i \leq v_{i,\mathrm{max}}$). It leads to the definition of a set of nonnegative generating vectors. The differences between linear and convex bases are summarized in Table 9.1. An important feature of the convex basis is that it is *unique* and determined based on network topology. However, the number of convex basis vectors can be *greater* than the dimension of the null space, leading to multiple ways to represent a flux distribution with the convex basis vectors.

Finite or closed spaces

The elementary reactions have nonnegative fluxes, $v_i \geq 0$. In addition, they have an upper bound, $v_i \leq v_{i,\mathrm{max}}$. Thus, the allowable flux vectors are in a rectangular hyperbox in the positive orthant of the flux space, bounded by planes parallel to each axis as defined by $v_{i,\mathrm{max}}$ (see Figure 9.5). This

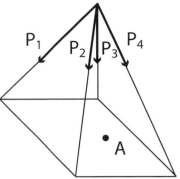

Figure 9.3: A four-sided pyramid. This object is a bounded 3D convex space. There are four convex basis vectors. Any nonnegative combination of these four vectors will generate a point, A, inside the pyramid.

hyperbox contains all allowable flux states, both steady-state and dynamic. The flux balance equation $\mathbf{Sv} = 0$ is a hyperplane that intersects the hyperbox forming a finite segment of a hyperplane. This intersection is a polytope in which all the steady-state flux distributions lie. This polytope can be spanned by the convex basis vectors that are *edges* of the polytope (see Figure 9.5C) with restricted ranges on the weights. In other words,

$$\mathbf{v}_{ss} = \sum_{k=1} \alpha_k \mathbf{p}_k \qquad (9.8)$$

where $\mathbf{p}_k$ are the edges, or extreme states, and α_k are the weights that are positive and bounded, $0 \leq \alpha_{i,\min} \leq \alpha_i \leq \alpha_{i,\max}$. The $\mathbf{p}_k$ are the so-called extreme pathways and are discussed in the following section.

Illustrative examples

Let us consider the consequences of the nonnegativity properties of the fluxes and the flux balance equations. Unfortunately, only the simplest of cases can be viewed in 3D, and we must use conceptual illustrations.

- A simple illustration of a 3D convex object is a four-sided pyramid, seen in Figure 9.3. As illustrated in the figure, this object is described by four convex basis vectors. A nonnegative combination of these four vectors leads to the generation of a point inside the pyramid.
- The flux vector lies in the positive orthant further constrained by the flux balance equations (Figure 9.4). The general shape of this space is one of a cone or a triangle in a 2D representation. Two pairs of nonnegative basis vectors are shown. Only the vectors that lie on the edges can give a nonnegative representation of all the points in the cone. It is thus a unique convex basis. Note that to represent point A in Figure 9.4, we need to use a negative weighting on the dashed set of basis vectors.

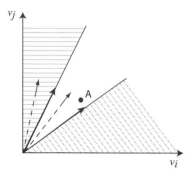

Figure 9.4: Schematic illustration of a flux cone in the positive orthant. The dashed vectors are non-negative basis vectors, while the pair of solid vectors are a convex basis, with which the entire cone can be spanned with $\alpha_i > 0$. Note that representing the point A would require a negative weighting on the dashed set of basis vectors.

The simple flux split

A node with three reaction links forms a simple flux split (Figure 9.5(A). The min, max constraints on the reactions form a 3D box that is intercepted by the plane formed by the flux balance

$$0 = v_3 - v_1 - v_2 = \langle (-1, -1, 1) \cdot (v_1, v_2, v_3) \rangle \quad (9.9)$$

forming a segment of a plane (Figure 9.5(B)). This 2D polytope is spanned by two convex basis vectors (Figure 9.5(C)). These basis vectors are $\mathbf{b}_1 = (v_1, 0, v_3)$ and $\mathbf{b}_2 = (0, v_2, v_3)$. The length of these basis vectors is thus limited by the $v_{i,\max}$ values. The relative magnitudes of the $v_{i,\max}$ values lead to *dominant* and *redundant* constraints, as illustrated in Figure 9.6. The

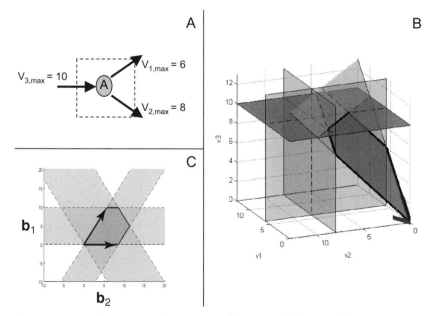

Figure 9.5: Forming a 2D polytope in 3D. (A) A simple flux split. (B) The 2D null space is constrained by the $v_{\max}$ planes corresponding to the three reactions in the network. (C) A 2D representation of the finite null space. Modified from [179].

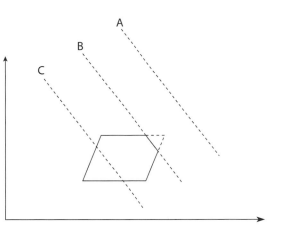

Figure 9.6: Bounding the flux cone; redundant and dominant constraints. Three constraints on $v_{3,max}$ in Figure 9.5 are illustrated; A: $v_{3,max} > 14(=v_{2,max} + v_{1,max})$; B: $6 < v_{3,max} < 14$; C: $v_{3,max} < 6(= \min\{v_{1,max}, v_{2,max}\})$. In case A, $v_{3,max}$ is redundant; in case B, all $v_{i,max}$ are relevant; and in case C, $v_{3,max}$ is relevant, while $v_{1,max}$ and $v_{2,max}$ are redundant.

upper bounds on the fluxes close the flux cone to form a polytope. This finite set has *extreme points* that can be used to define the space.

Varying constraints and biological interpretation

The constraints offer a mechanism to determine the effects of various parameters on the achievable functional states of a network. For instance, enzymopathies can reduce the maximum flux through a particular reaction in a network. Such constraints could reduce the numerical value for $\alpha_{i,max}$ and thus shorten the maximum length of an extreme pathway vector (see Figure 9.7). In such a case, several functional states are not possible. A desirable functional state may thus be eliminated, possibly leading to a pathological condition [103]. The consequences of complete gene deletions also can be analyzed in this fashion (see Chapter 16).

Some key concepts: Mathematics versus biology

The finite null space contains all the allowable steady-state flux distributions through the network. It provides a nice link between mathematical and biological concepts:

- The null space represents all the possible functional, or phenotypic, states of a network.
- A particular point in the polytope represents one network function or one particular phenotypic state.

Figure 9.7: Changing constraints and shrinking polytopes. If a v_{max} constraint is reduced, a portion of the solution space may become inaccessible (shaded region). Enzymopathies can have this effect [103].

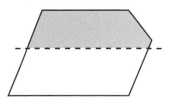

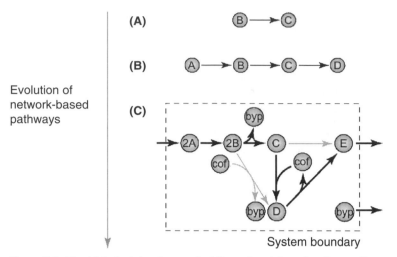

Evolution of
network-based
pathways

Figure 9.8: The historical development of the network-based pathway diagram: From reactions to pathways to networks. Taken from [163].

- As we will see in Chapter 16, there are equivalent points in the cone that lead to the same overall functional state of a network. Biologically, such conditions are called *silent phenotypes*.
- The edges of the flux cone are the unique extreme pathways. Any flux state in the cone can be decomposed into the extreme pathways. The unique set of extreme pathways thus gives a mathematical description of the range of flux levels that are allowed.

Perspective: From reactions to pathways

Early in the history of biochemistry, enzymes isolated from cells were shown to be able to carry out specific chemical reactions (Figure 9.8A). It was then recognized that the products of one reaction were substrates of another. Thus, one could link different chemical transformations to form a series of reactions (Figure 9.8B), to form basic metabolic pathways, such as glycolysis, the TCA cycle, and so on. The definition and biochemical functions of such pathways have been taught to generations of life scientists. With the advent of whole genome sequencing and the development of network reconstruction methods (see Part I), we can now piece together entire networks. The properties of such networks can and must be studied. In the words of Uwe Sauer, "Pathways are concepts, but networks are reality."

Interestingly, as illustrated earlier, the finite null space of the stoichiometric matrix has a natural set of basis vectors that can be used to span all allowable network states. Thus, network-based definitions of pathways

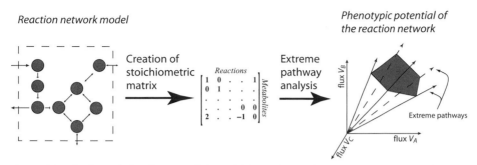

Figure 9.9: The bounded null space of the stoichiometric matrix as a convex cone. Redrawn from [163].

emerged (see Figure 9.8), that account for the function of the network as a whole. These pathways are mathematically defined and thus free of human bias in their definition and are useful for studying the systems properties of networks. Next, we discuss such pathways, and they are put to use in Chapter 13 to study large-scale networks.

9.3 Extreme Pathways

Biochemically meaningful steady-state flux solutions can be represented by a nonnegative linear combination of convex basis vectors as

$$\mathbf{v}_{ss} = \sum \alpha_i \mathbf{p}_i \quad \text{where } 0 \leq \alpha_i \leq \alpha_{i,\max} \tag{9.10}$$

The vectors $\mathbf{p}_i$ are a unique convex generating set, but α_i may not be unique for a given $\mathbf{v}_{ss}$. The number of the convex basis vectors can exceed $n - r$. The $\mathbf{p}_i$ have been extensively studied [159, 193]. They correspond to the edges of a cone in an $(n - r)$-dimensional space (see Figure 9.9). They correspond to pathways when represented on a flux map and are called *extreme pathways*, since they lie at the edges of the bounded null space in its conical representation.

The flux cone

For conceptual purposes, the extreme pathways can then be visualized as vectors in a high-dimensional space, whose axes correspond to the flux levels through the individual reactions (Figure 9.9). A numerical value on a given axis is the flux level in the corresponding reaction. The flux cone depicted in Figure 9.9 is the set of points that correspond to allowable flux values in the metabolic network. The cone can also be mathematically

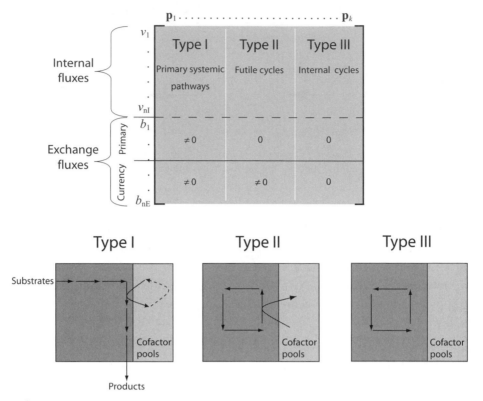

Figure 9.10: Classification of extreme pathways based on the exchange fluxes that they contain. (A) The structure of the pathway matrix **P** for the tree types of extreme pathways; (B) The corresponding flux maps. Redrawn from [174].

described as

$$C = \{\mathbf{v} : \mathbf{v} = \sum_{i=1}^{p} \alpha_i \mathbf{p}_i, \alpha_i \geq 0, \text{ for all } i\} \quad (9.11)$$

where C represents the convex cone encompassing all possible steady-state flux distributions in a metabolic network; $\mathbf{v}$ represents the vector of fluxes for each reaction in the network; p is the number of extreme pathways $(p \geq n - r)$, and α_i and $\mathbf{p}_i$ are the weights and the extreme pathways of the network, respectively.

Classification of the extreme pathways

Extreme pathways have been classified into three groups according to their use of exchange metabolites (Figure 9.10). The extreme pathways are computed based on the null space of $\mathbf{S}_{\text{exch}}$. The exchange fluxes are grouped into two categories: external fluxes and fluxes external to metabolism but internal to the cell. Thus, the first set of exchange fluxes are with the surroundings of the cell, while the second set represent a virtual boundary separating metabolism from other cellular functions. The second category primarily contains the use of *currency metabolites*, such as carrier and co-factor molecules. The *pathway matrix* **P** (whose columns are the $\mathbf{p}_i$) and a schematic of what it represents is shown in Figure 9.10.

With this definition of exchange fluxes, the extreme pathways are classified into three categories based on which exchange fluxes they contain.

- type I extreme pathways involve the conversion of primary inputs into primary outputs and thus contain exchange fluxes with the environment.
- type II extreme pathways involve the internal exchange of currency (carrier) metabolites only, such as ATP and NADH.
- type III extreme pathways are solely internal cycles; there is zero flux across all system boundaries.

By either including currency metabolites as internal to the system or considering currency metabolites as primary exchange metabolites, an extreme pathway analysis yields zero type II pathways. It has been demonstrated that type III pathways are thermodynamically infeasible and consequently the corresponding net fluxes through them are zero [174].

Simple reactions

Extreme pathways can be computed for simple reactions (Figure 9.11). If they operate in a closed system, the only pathway is a type III that corresponds to the two elementary reactions operating in the opposite direction. This corresponds to chemical equilibria when the fluxes have to be equal and opposite. Thus, there is no net flux associated with the type III pathway.

If the simple reactions operate in an open setting, type I extreme pathways appear (see Figure 9.11). For a simple reversible reaction, a straight through type I pathway occurs in addition to the type III pathway that is also present in the closed system. Adding exchange reactions as columns in **S** increases the dimension of the null space. This linear type I pathway represents the common notion of a metabolic pathway.

Most reactions in cells involve carrier molecules. The carrier coupled reaction leads to exchange fluxes within the cell (see Figure 9.11). The type I pathways that arise describe not only the entry and exit of the primary substrate but also the charging of a carrier molecule, with the uncharged form entering and the charged form leaving (or vice versa). This type I pathway thus describes the balancing of all network functions at the same time. This type of an extreme pathway will be typical of what is found in real networks and is of greatest interest.

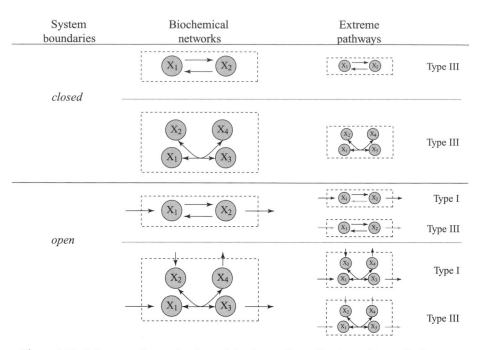

Figure 9.11: Extreme pathway structure of simple reactions. Courtesy of Jason Papin.

Skeleton metabolic pathways

The extreme pathway structure of simple reactions is not complicated. As the number of reactions that occur simultaneously goes up, the extreme pathway structure becomes more complex. A skeleton pathway is shown in Figure 9.12, and it represents the basic structure of substrate-level phosphorylation in glycolysis. The substrate enters and is "charged" with a phosphate group by a carrier molecule, i.e., ATP. Then, after a chemical transformation, two charged carrier molecules are produced. Note that for this network to be strictly elementally balanced, x_1 must contain the second phosphate group that the carrier molecule takes away. The charging reaction is not reversible but can proceed in the opposite reaction by the splitting of the phosphate group. Either x_4 or x_5 can leave the system, as with pyruvate and lactate in glycolysis.

This simple skeleton representation of glycolysis has five extreme pathways. There are two type I pathways that result in the secretion of x_4 and x_5 respectively, and the net production of one charged cofactor molecule. There is one type II pathway that represents the dissipation of the high-energy phosphate bond, a futile cycle. This pathway can operate in only one direction. Then there are two type III pathways that will have no net flux.

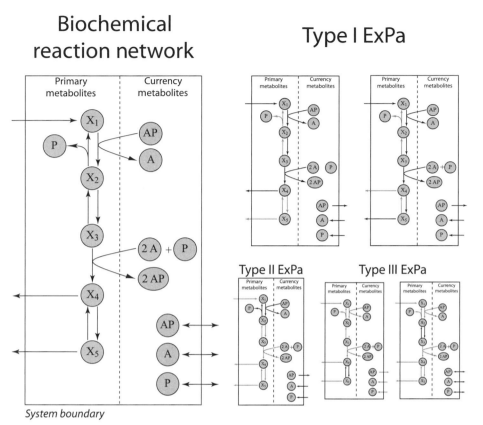

Figure 9.12: Extreme pathway structure of a skeleton phosphate metabolism in glycolysis. Courtesy of Jason Papin.

Large-scale networks

We defer the analysis of large-scale networks to Chapter 13. Although the discussion in this section is mostly focused on metabolic networks, it should be clear to the reader that this analysis applies to any network whose reactions have been defined and stoichiometrically described. For instance, the JAK-STAT signaling network in B lymphocytes has been stoichiometrically represented and its extreme pathways analyzed [160].

Computing extreme pathways

Algorithms have been developed to compute the extreme pathways [190]. The details of such algorithms are not important here, but open source code has been provided to compute them [13]. The computation of extreme pathways for small systems is relatively easy. However, as the size of a network grows, the number of extreme pathways grows much faster. The combinatorial nature of their computation leads to an N-P hard problem. Thus,

---

Table 9.2: The history of the use of convex analysis of steady-state flux vectors. From [159].

1980	Stoichiometric network analysis developed to study instability in inorganic chemical reaction networks. Relied on kinetic information and utilized concepts of convex analysis [31].
1988	Artificial intelligence algorithm developed to search through reaction networks for the identification/synthesis of biochemical pathways [204].
1990	Stoichiometric constraints used to synthesize pathways again using artificial intelligence [135].
1992	Development of concept implementing null space vectors as component pathways [59].
1994	Convex analysis first applied to metabolic networks with the introduction of a nonunique set of elementary modes [199].
1996	Pathway analysis used to optimize bacterial strain design for highly efficient production of aromatic amino acid precursors [121].
1999	(*1996–1999*) Software for elementary mode analysis. Development of algorithm to use elementary modes to calculate enzyme subsets [166].
	Elementary modes used for comparative genome analysis [42].
2000	Extreme pathways, the unique and irreducible set of elementary modes, are developed and applied to the study of large-scale metabolic networks [190].
2002	Extreme pathways are calculated for genome-scale metabolic networks [162, 174].
	Elementary modes are determined for a recombinant strain of yeast [28].
	Gene expression is correlated to enzyme subsets in yeast calculated from elementary modes [200].
	Elementary modes of a recombinant strain of *M. estorquens* AM1 are analyzed [226].
	Regulatory logic is incorporated into extreme pathway analysis [35].
	Human red blood cell metabolism is analyzed with extreme pathways [246].

---

it has been possible only to compute extreme pathways at the genome-scale for small matrices and under a limited set of conditions [162, 178]. The development of robust algorithms for their computation is needed and so are analysis methods to interpret the results. In general, a full set of extreme pathways may not be of importance, but a subset of them will be. Fortunately, linear programing can be used to find particular extreme pathways of interest, which typically is a very fast computation (see Chapter 15).

History of convex pathway vectors

The use of convex analysis of flux solution spaces has a history that spans more than 20 years. The highlights of these developments are shown in Table 9.2. The original *extreme currents* by B. Clarke in 1980 and the *elementary modes* by S. Schuster are based on a similar definition as extreme pathways and are related quantities. These definitions differ in the way the reactions are represented. For instance, the elementary modes combine the two irreversible elementary reactions, representing a reversible

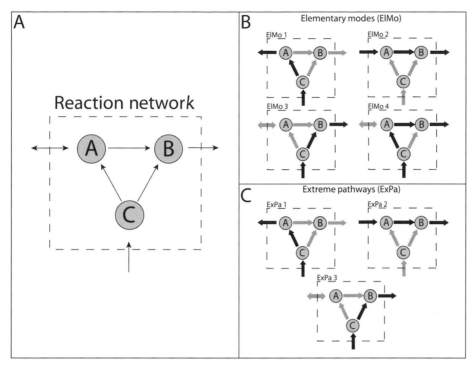

Figure 9.13: Simple example of a biochemical network with corresponding extreme pathways (ExPa) and elementary modes (ElMo). (A) The network consists of three metabolites, three internal reactions, and three exchange reactions. There are four elementary modes (B) and three extreme pathways (C) in this network. The difference between the two sets of pathways revolves around the use of the reversible exchange flux for the metabolite A. ElMo 4 is a nonnegative linear combination of ExPa 1 and ExPa 2, since the reversible exchange flux for metabolite A can be canceled out. From [161].

chemical conversion into one net reaction that can take on positive and negative values. The choice of how the reactions are represented leads to different forms of a polytope, since the axes of the space differ.

Contrasting elementary modes and extreme pathways

These two types of network-based pathway definitions have received much attention. The extreme pathways and elementary modes for a simple reaction system are illustrated in Figure 9.13. There may be fewer extreme pathways than elementary modes. Thus, elementary modes are a superset of the extreme pathways. The primary difference between the two definitions is the representation of exchange fluxes. If the exchange fluxes are all irreversible, the extreme pathways and elementary modes are equivalent. If the exchange fluxes are reversible, there are more elementary modes than extreme pathways. For instance, for the metabolic network of the human red blood cell, there are 55 extreme pathways but 6,180 elementary modes [158]. This larger number has it roots in the same feature as displayed in Figure 9.13. The extreme pathways are a convex basis set. Thus, once added up, a reversible exchange flux can disappear (as in the representation of elementary mode 4 in Figure 9.13).

9.4   Summary

➤ The stoichiometric matrix has a null space that corresponds to a linear combination of the reaction vectors that add up to zero; so-called link-neutral combinations.

➤ The orthonormal basis given by SVD does not yield a useful biochemical interpretation of the null space of the stoichiometric matrix.

➤ Convex basis vectors for the null space can be formulated by considering elemental reactions only. Since they are irreversible ($0 \leq v_i$) and finite in magnitude ($v_i \leq v_{i,\max}$), it only makes sense to add them in a positive fashion leading to a convex representation of the null space.

➤ The convex representation has edges that represent a set of vectors that span the convex space in a nonnegative fashion.

➤ These edges correspond to biochemical pathways and are extreme functions that a network can have; therefore they are called *extreme pathways*.

➤ The extreme pathways are classified based on the exchange reactions that they contain. Type I involve the transfer of properties to currency molecules (carriers or cofactors), type II have no external exchange reaction and represent irreversible futile cycles, and type III have no exchange reactions and by thermodynamic restrictions must have zero flux.

➤ The number of extreme pathways grows faster than the number of components in a network, giving rise to the need to study them in large numbers.

9.5   Further Reading

Clarke, B.L., "Stoichiometric network analysis," *Cell Biophysics*, **12**:237–253 (1988).

Clarke, B.L., "Stability of complex reaction networks," in ADVANCES IN CHEMICAL PHYSICS (I. Prigogine and S.A., Rice, Eds.), pp. 1–215, Wiley, New York (1980).

Heinrich, R., and Schuster, S., THE REGULATION OF CELLULAR SYSTEMS, Chapman and Hall, New York (1996).

Papin, J.A., Price, N.D., Wiback, S.J., Fell, D.A., and Palsson, B., "Metabolic pathways in the post-genome era," *Trends in Biochemical Sciences,* **28**:250–258 (2003).

Papin, J.A., Stelling, J., Price, N.D., Klamt, S., Schuster, S., and Palsson, B.O., "Comparison of network-based pathway analysis methods," *Trends in Biotechnology,* **22**:400–405 (2004).

Rockafellar, R. T., CONVEX ANALYSIS, Princeton University Press, Princeton, NJ (1970).

Schilling, C.H., Schuster, S., Palsson, B.O., and Heinrich, R., "Metabolic pathway analysis: Basic concepts and scientific applications in the post-genomic era," *Biotechnology Progress,* **15**:296–303 (1999).

Schuster, S., and Hilgetag, C., "On elementary flux modes in biochemical reaction systems at steady-state," *Journal of Biological Systems,* **2**:165–182 (1994).

Wiback, S.J., and Palsson, B.O., "Extreme pathway analysis of human red blood cell metabolism," *Biophysical Journal,* **83**(2):808–818 (2002).

CHAPTER 10

The Left Null Space of **S**

The left null space of the stoichiometric matrix contains combinations of the time derivatives that add up to zero. This summation adds up to zero in both dynamic and steady states of the network; hence it contains *time-invariant* quantities. This condition is also called *node neutrality*. As with the (right) null space, the choice of basis for the left null space is important in describing its contents in biochemically and biologically meaningful terms. A convex representation of the left null space has proven useful.

10.1 Definition

The left null space of **S** is defined by

$$\mathbf{LS} = \mathbf{0} \tag{10.1}$$

The dimension of the left null space of the stoichiometric matrix is $m - r$. Equation 10.1 represents a linear combination of nodes whose links sum to zero, as can be seen pictorially from

$$(\equiv)(|||) = \mathbf{0} \tag{10.2}$$

where the vertical lines represent column vectors and the horizontal lines represent row vectors. Thus the rows of **L** are a set of linearly independent vectors, $\mathbf{l}_i, i \in [1, m - r]$, spanning the left null space, and they are orthogonal to the reaction vectors ($\mathbf{s}_j$) comprising **S**

$$\langle \mathbf{l}_i \cdot \mathbf{s}_j \rangle = 0 \tag{10.3}$$

The set of $\mathbf{l}_i$'s may represent mass conservation of atomic elements (e.g., hydrogen, carbon, oxygen), molecular subunits (e.g., carboxyl group, hydroxyl group), or chemical moieties (e.g., pentose, purine), as discussed

Chapter 6. The elemental vectors $\mathbf{e}_i$, for instance, represent conservation relationships for the elements (see equation 6.19). Thus, the vectors that span the left null space represent the conservation relationships that are orthogonal to the reaction vectors ($\mathbf{s}_j$) that span the column space.

10.2 The Time Invariants

The conservation relationships

One can use equation 6.3 to show that equation 10.1 represents

$$\mathbf{l}_i \frac{d\mathbf{x}}{dt} = 0 \tag{10.4}$$

which is a *conservation relationship*, or a summation of concentrations, called a *pool*, that is time invariant. All the time derivatives can be written simultaneously as

$$\frac{d}{dt}(\mathbf{Lx}) = \mathbf{0} \tag{10.5}$$

which represents all the conservation relationships and defines the pools. While there can be dynamic motion taking place in the column space along the reaction vectors, these motions do not change the total amount of mass in the time-invariant pools (see Chapter 11 for examples). Note that since the basis for the left null space is nonunique, there are many ways to represent these metabolic pools. The time derivatives can be positive or negative.

Pool sizes

Equation 10.5 can be integrated to give the *mass conservation equations*:

$$\mathbf{Lx} = \mathbf{a} \tag{10.6}$$

where $\mathbf{a}$ is a vector that gives the sizes (the total concentration) of the pools. Note that equation 10.6 defines an *affine* hyperplane, that is, a plane that does not go through the origin. This hyperplane is the *concentration space* in which the concentration vector $\mathbf{x}$ resides. Since the concentrations x_i are nonnegative quantities, like fluxes through elementary reactions, a convex representation of the concentration space is useful. The convex basis for the left null space can be computed in the same way as the (right) null space by transposing $\mathbf{S}$.

As discussed in Chapter 6, a compound map can be formed based on the negative transpose of the stoichiometric matrix. In Chapter 9 we introduced extreme pathways and graphed them as vectors on flux maps.

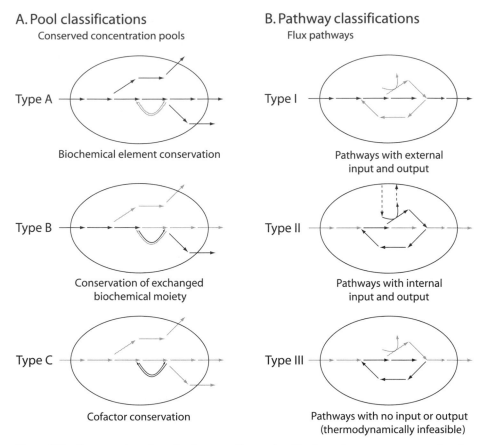

A. Pool classifications
Conserved concentration pools

Type A

Biochemical element conservation

Type B

Conservation of exchanged
biochemical moiety

Type C

Cofactor conservation

B. Pathway classifications
Flux pathways

Type I

Pathways with external
input and output

Type II

Pathways with internal
input and output

Type III

Pathways with no input or output
(thermodynamically infeasible)

Figure 10.1: Conserved pools and extreme pathway classifications shown on compound and reaction maps, respectively. (A) Type A, B, and C metabolic pools correspond to the conservation of biochemical elements, metabolic moieties common to the primary and secondary metabolites, and cofactor conservation, respectively. (B) Type I, II, and III extreme pathways correspond to through pathways, futile cycles coupled to cofactor utilization, and internal loops that are thermodynamically infeasible [174], respectively. Prepared by Iman Famili.

In a similar fashion, the conservation quantities represent groups of arrows on a compound map. These compound maps are useful to display the metabolic pools.

Classifying the pools

Extreme pathway analysis led to the classification of three basic types of convex basis vectors (see Chapter 9). These three categories correspond to "through" flux pathways (type I), "futile cycles" coupled to cofactor use (type II), and "internal" cycles (type III). In an analogous fashion, the conservation quantities can be grouped into three basic types: types A, B, and C (Figure 10.1). The classification is based on grouping the compounds into *primary* and *secondary* categories (Figure 10.2). The primary

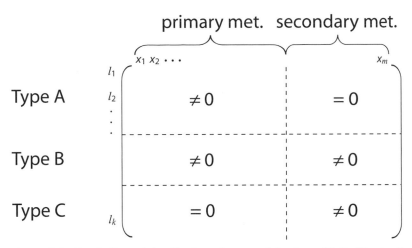

Figure 10.2: Metabolic pool classification schema and structure of **L**. Partitioning the metabolites into *primary* and *secondary* allows for the classification of metabolic pools. In the absence of the secondary and primary metabolite participation in conservation pools, the vectors are classified as types A and C, respectively. The remaining vectors are grouped as type B pools with both metabolite types present. Note the similarity to Figure 9.10. Prepared by Iman Famili.

compounds contain the primary molecule structures that are undergoing serial transformations (e.g., the carbon backbone of glucose in glycolysis). The secondary compounds function as currency molecules (carriers and cofactors) and generally remain internal to the cell (e.g., ATP). Based on this classification, the three types of conservation pools correspond to:

- type A pools that are composed only of the primary compounds;
- type B pools that contain both primary and secondary compounds internal to the system; and
- type C pools that comprise only of secondary compounds.

type B pools generally represent the conserved moieties (or currencies) that are exchanged from one compound to another, such as a hydroxyl or a phosphate group.

Reference states
We note that the conservation equations (equation 10.6) hold for all states of the network, and thus

$$\mathbf{Lx} = \mathbf{Lx}_{\text{ref}} = \mathbf{a} \tag{10.7}$$

or

$$\mathbf{L}(\mathbf{x} - \mathbf{x}_{\text{ref}}) = \mathbf{0} \tag{10.8}$$

The initial or steady-state conditions can be used to set the pool sizes, a_i. Note that **x** is *not* orthogonal to the left null space of **S**, whereas $d\mathbf{x}/dt$ and $(\mathbf{x} - \mathbf{x}_{\text{ref}})$ are.

A reference state for **x** can be selected. Equation 10.8 is a parameterization of the concentration space. If we select $\mathbf{x}_{\text{ref}}$ that is parallel to the row vectors of **L**, we can span the affine concentration space using the reaction vectors $\mathbf{s}_i$, since they are an orthogonal set of vectors. Thus, we can state two criteria for such a selection:

1. $\mathbf{x}_{\text{ref}}$ is orthogonal to $\mathbf{s}_i$ (since $\mathbf{x}_{\text{ref}}$ is in the left null space).
2. $\mathbf{x} - \mathbf{x}_{\text{ref}}$ is orthogonal to $\mathbf{l}_i$ (follows from equation 10.8).

These criteria are illustrated in Figure 10.3. These two conditions are linear and together lead to a unique solution for $\mathbf{x}_{\text{ref}}$. Thus, we can convert the affine representation of the concentration space (equation 10.6) into a non-affine representation using $\mathbf{x} - \mathbf{x}_{\text{ref}}$ and use the same basis for this space as the left null space of **S**.

10.3 Single Reactions and Pool Formation

Simple reversible reaction

Consider the simple reaction of equation 6.11. The stoichiometric matrix and the convex basis for the left null space are

$$\mathbf{S} = \begin{pmatrix} -1 & 1 \\ 1 & -1 \end{pmatrix} \quad \text{and} \quad \mathbf{L} = \mathbf{l}_1 = (1, 1) \tag{10.9}$$

If $a_1 = 1$, then a parameterization of **x** is given by

$$\mathbf{x} = \begin{pmatrix} 1 \\ 0 \end{pmatrix} + \xi \begin{pmatrix} -1 \\ 1 \end{pmatrix} \quad \xi \in [0, 1] \tag{10.10}$$

We can now determine $\mathbf{x}_{\text{ref}}$. The first criteria is given by

$$(-1, 1) \cdot \begin{pmatrix} x_{1,\text{ref}} \\ x_{2,\text{ref}} \end{pmatrix} = 0 \tag{10.11}$$

and thus $x_{1,\text{ref}} = x_{2,\text{ref}}$. The second criteria is

$$(1, 1) \cdot \begin{pmatrix} x_1 - x_{1,\text{ref}} \\ x_2 - x_{2,\text{ref}} \end{pmatrix} = 0 \tag{10.12}$$

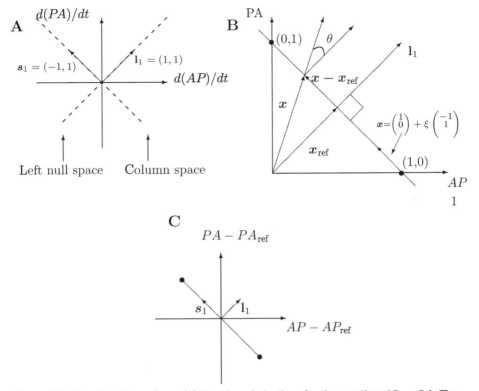

Figure 10.3: The definition of $\mathbf{x}_{\text{ref}}$. (A) The time derivatives for the reaction $AP \rightleftharpoons PA$. The vectors $\mathbf{l}_1$ and $\mathbf{s}_1$ that span the left null and column space, respectively, are shown. B) The concentration space for the same reaction when the total concentration $AP + PA = a_1 = 1$. The starting point of $(1, 0)$ and the parametrization of the vector $\mathbf{x}$ is shown. (C) The subtraction of the reference concentration state $(AP_{\text{ref}}, PA_{\text{ref}})$ to form a homogeneous concentration space that is spaced by $\mathbf{s}_1$ and is orthogonal to $\mathbf{l}_1$. Note that the concentration space is closed and has end points that represent the extreme concentration states of this reaction.

The solution of these two equations is

$$\begin{pmatrix} x_{1,\text{ref}} \\ x_{2,\text{ref}} \end{pmatrix} = \begin{pmatrix} 1/2 \\ 1/2 \end{pmatrix} \tag{10.13}$$

since $x_1 + x_2 = a_1 = 1$. Thus, we finally get the parameterization of the concentration vector as

$$\mathbf{x} - \mathbf{x}_{\text{ref}} = \begin{pmatrix} 1/2 \\ -1/2 \end{pmatrix} + \xi \begin{pmatrix} -1 \\ 1 \end{pmatrix} \quad \xi \in [0, 1] \tag{10.14}$$

and $\mathbf{x} - \mathbf{x}_{\text{ref}}$ is now spanned by $\mathbf{s}_1 = (-1, 1)^{\text{T}}$. Figure 10.3C shows how this representation of the concentration space goes through the origin, is spanned by $\mathbf{s}_1$, and has extreme states (the end points). Thus, the concentration space is *bounded*.

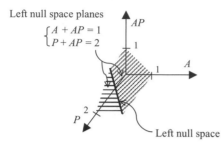

Left null space planes
$$\begin{cases} A + AP = 1 \\ P + AP = 2 \end{cases}$$

Figure 10.4: The one-dimensional concentration space for the reaction $A + P \rightarrow AP$, where the total concentrations are $A + AP = 1$ and $P + AP = 2$. The vectors $\mathbf{l}_1$ and $\mathbf{l}_2$ that span the left null space are shown.

Bilinear association

Next, consider the reaction in equation 6.13. The stoichiometric matrix and the convex basis for the left null space are (if the concentrations in $\mathbf{x}$ are ordered as A, P, AP)

$$\mathbf{S} = \begin{pmatrix} -1 \\ -1 \\ 1 \end{pmatrix}, \quad \mathbf{L} = \begin{pmatrix} 1 & 0 & 1 \\ 0 & 1 & 1 \end{pmatrix} \tag{10.15}$$

If $\mathbf{a} = (1, 2)^{\mathrm{T}}$, then the one-dimensional concentration space is given by the parameterization (see Figure 10.4):

$$\mathbf{x} = \begin{pmatrix} 1 \\ 2 \\ 0 \end{pmatrix} + \xi \begin{pmatrix} -1 \\ -1 \\ 1 \end{pmatrix}, \quad \xi \in [0, 1] \tag{10.16}$$

$\mathbf{x}_{\mathrm{ref}}$ can be determined from $\mathbf{L}(\mathbf{x} - \mathbf{x}_{\mathrm{ref}}) = 0$ and $\langle \mathbf{s}_1 \cdot \mathbf{x}_{\mathrm{ref}} \rangle = 0$ to obtain $\mathbf{x}_{\mathrm{ref}} = (0, 1, 1)^{\mathrm{T}}$. This reference state then leads to the parameterization of the concentration space as

$$\mathbf{x} - \mathbf{x}_{\mathrm{ref}} = \begin{pmatrix} 1 \\ 1 \\ -1 \end{pmatrix} + \xi \begin{pmatrix} -1 \\ -1 \\ 1 \end{pmatrix} \quad \xi \in [0, 1] \tag{10.17}$$

As shown in Figure 10.5, this forms a line between $(1, 1, -1)$ and $(0, 0, 0)$ in the space formed by the shifted concentration space. It is orthogonal to **L** and is spanned by $\mathbf{s}_1$.

Carrier-coupled reaction

Finally consider the reaction in equation 6.15. The stoichiometric matrix and the convex basis for the left null space is given by (the entries of $\mathbf{x}$ are

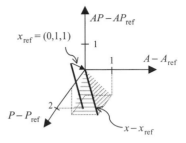

Figure 10.5: The affine concentration space for the reaction $A + P \rightarrow AP$, where the total concentrations are $A + AP = 1$ and $P + AP = 2$. The vectors $\mathbf{l}_1$ and $\mathbf{l}_2$ that span the left null space are shown.

ordered by (CP, C, AP, A))

$$S = \begin{pmatrix} 1 \\ -1 \\ -1 \\ 1 \end{pmatrix}, \quad L = \begin{pmatrix} 1 & 1 & 0 & 0 \\ 0 & 0 & 1 & 1 \\ 1 & 0 & 1 & 0 \\ - & - & - & - \\ 0 & 1 & 0 & 1 \end{pmatrix} \qquad (10.18)$$

The left null space is three dimensional, but its convex basis has four row vectors as shown.

- The first pool is a conservation of the primary substrate pool $C \ (= C + CP)$ and is a type A pool.
- The second pool is a conservation of the cofactor $A \ (= A + AP)$ and is a type C pool.
- The third pool is a conservation of the phosphorylated compounds $(= CP + AP)$ and represents the total energy inventory, or *occupancy*, in the system.
- The last pool is a *vacancy* pool $(C + A)$ that represents the low-energy state of the participating compounds.

The latter two pools relate to exchange of a property (energy in the form of a high-energy phosphate bond in this case) and form a *conjugate pair* of occupied and vacant carrier states.

If $\mathbf{a} = (1, 2, 2, 1)^T$, then the four convex vectors can be depicted on a two-dimensional graph as shown in Figure 10.6. By considering $L(\mathbf{x} - \mathbf{x}_{ref}) = 0$ and $< \mathbf{x}_{ref} \cdot \mathbf{s}_1 >= 0$, we get $\mathbf{x}_{ref} = (3/2, -1/2, 1/2, 3/2)$, which leads to the parameterization

$$\mathbf{x} - \mathbf{x}_{ref} = \begin{pmatrix} 3/4 \\ 3/4 \\ 3/4 \\ 3/4 \end{pmatrix} + \xi \begin{pmatrix} 1 \\ -1 \\ -1 \\ 1 \end{pmatrix}, \quad \xi \in [0, 1] \qquad (10.19)$$

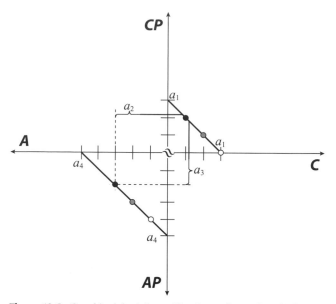

Figure 10.6: Graphical depiction of the three-dimensional left null space for the carrier-coupled reaction. The carrier-coupled reaction contains four convex conservation relationships as shown in Figure 9.9. The relative distance between the carbon and carrier metabolites remains constant and is determined by the magnitude of the conservation quantities, and so $C + CP = a_1$, $C + A = a_2$, $CP + AP = a_3$, and $A + AP = a_4$. The points with the same shading (black, grey, white) correspond to an identical state depicted in two distinct two-dimensional spaces. For example, the concentration state in which $C = 2$, $CP = 1$, $A = 2$, and $AP = 3$ is depicted by the grey point and satisfies $a_1 = 3$, $a_2 = 4$, $a_3 = 4$, and $a_4 = 5$. The states represented by the white and black points also satisfy theses pool sizes. The concentration solution space is the solid line shown in the two spaces. The symbol $\approx$ indicates that the origin is not common between the two two-dimensional spaces. Prepared by Iman Famili.

The exchange of the carried P moiety between C and A creates a conjugate pair of convex basis vectors in **L**. They correspond to the total P bound to AP and CP, or the P-occupancy in the system, and the vacant binding sites for P, or $C + A$.

Redox carrier coupled reactions
The coupling of the NADH redox carrier to a reaction is given by the chemical equation

$$RH_2 + NAD^+ \overset{v_1}{\underset{v_2}{\rightleftharpoons}} R + NADH + H^+ \tag{10.20}$$

where two protons on a moiety R are captured by the oxidized form of NAD releasing R and a free proton. The stoichiometric matrix for this system is

$$
\mathbf{S} = \begin{pmatrix} -1 & 1 \\ -1 & 1 \\ 1 & -1 \\ 1 & -1 \\ 1 & -1 \end{pmatrix}
\tag{10.21}
$$

The convex basis for the left null space of this stoichiometric matrix is

RH_2	NAD^+	R	H^+	$NADH$	Type	Pool interpretation
1	0	1	0	0	A	Total R
1	0	0	0	1	B	Redox occupancy 1
1	0	0	1	0	B	Redox occupancy 2
0	1	1	0	0	B	Redox vacancy
0	1	0	0	1	C	Total redox carrier 1
0	1	0	1	0	C	Total redox carrier 2

The reaction map and the compound map are shown in Figure 10.7. The pools are readily interpreted as indicated. There are three conservation pools of primary and secondary metabolites (types A and C). The first is associated with the substrate, and then two are associated with the redox carriers. The origin of the latter two is similar to that in the bimolecular association treated earlier. There are three exchange property conservation pools (type B). Two are total "occupancy" pools of redox equivalents, which represent total concentration of a reduced form of compounds, and one is a "vacancy" pool that represents the conjugate oxidized form of the pair.

10.4 Multiple Reactions and Pool Formation

The pool formation in single elementary reactions begins to show the usefulness of convex basis vectors to generate chemically meaningful definitions of pools. We now look at more complicated cases that enable us to begin to scale up the use of these definitions to larger networks. The time invariants derived for the elementary reactions can be displayed on a compound map (see the first three cases in Figure 10.8). These pools, once plotted on the compound map, look very much like pathways on a flux map, and the quantities being conserved are clearly displayed on such maps. We now pursue this representation for coupled reactions.

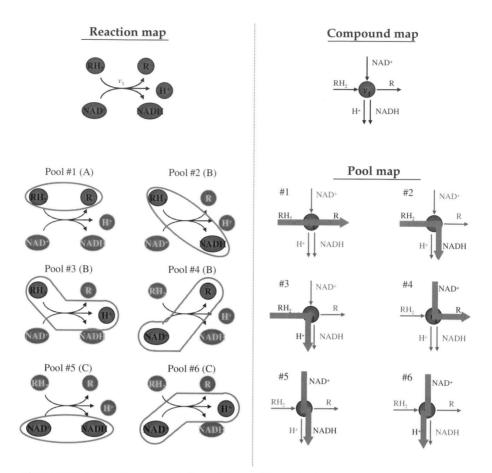

Figure 10.7: Schematic representation of the pool formation in redox coupled reaction equation 10.20. Prepared by Iman Famili.

Combining elementary reactions

As one combines elementary reactions, the pools become more interesting. Case IV in Figure 10.8 shows a combination of two serial carrier coupled reactions. Combining these two reactions leads to conservation quantities that are similar to the single carrier coupled reaction. Total substrate $(C + CP + CP_2)$ and carrier $(A + AP)$ pools appear, as well as P occupied $(CP + 2CP_2 + AP)$ and P vacant $(2C + CP + A)$ pools.

Case V in Figure 10.8 shows a cofactor coupled reaction followed by a condensation reaction where inorganic P is incorporated. Note that we still get a conservation of substrate (primary metabolite) and carrier (secondary metabolite). However, in this case there are two distinct types of P containing pools: one containing P that was always bound to a carrier and a new conservation quantity that corresponds to the free and incorporated P.

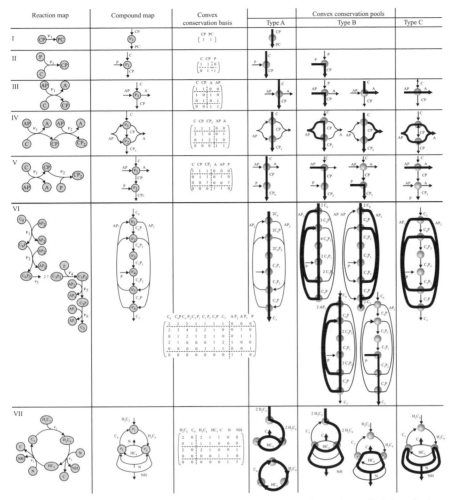

Figure 10.8: Pool maps for time-invariant quantities for simple and coupled chemical reactions. From [56].

Multiple redox coupled reactions

Biochemical reaction networks comprise multiple interacting redox or energy coupled reactions, as the following example chemical equations indicate:

$$RH_2 + NAD^+ \underset{v_2}{\overset{v_1}{\rightleftharpoons}} R + NADH + H^+ \tag{10.22}$$

as with the redox carrier coupled reaction in the previous section, followed by a chemical transformation of R

$$R \underset{v_4}{\overset{v_3}{\rightleftharpoons}} R' \tag{10.23}$$

followed by a reduction reaction reusing the redox carrier

$$R' + NADH + H^+ \underset{v_6}{\overset{v_5}{\rightleftharpoons}} R'H_2 + NAD^+ \qquad (10.24)$$

The use of NAD in glycolysis follows this basic reaction structure. The convex basis for the left null space is represented by

RH_2	NAD^+	R	H^+	$NADH$	R'	$R'H_2$	Type	Pool interpretation
1	0	1	0	0	1	1	A	Total R
1	0	0	0	1	0	1	B	Redox occupancy 1
	0	0	1	0	0	1	B	Redox occupancy 1
0	1	1	0	0	1	0	B	Redox vacancy
0	1	0	0	1	0	0	C	Total redox carrier 1
0	1	0	1	0	0	0	C	Total redox carrier 2

The reaction map and the compound map are shown in Figure 10.9. The pools are readily interpreted as in the redox carrier coupled reaction in the previous section. There are three conservation pools of primary and secondary metabolites (types A and C). The first is associated with the substrate and then two are associated with the redox carriers as in the previous example. There are also three exchange property conservation pools of type B. There are two that are total redox occupancy pools, and there is one "vacancy" pool. The latter three represent two coupled conjugate pairs of high and low redox states.

10.5 Pool Formation in Classical Pathways

The basic concepts outlined in the previous sections can be used to understand how pools form in realistic biological networks. Note that the symbols used in the reaction schemas that follow are not meant to directly indicate chemical elements.

Simplified glycolysis
A few additions to schema IV will give the skeletal structure of glycolysis (Figure 10.8). The convex basis for the left null space of the corresponding stoichiometric matrix has six conservation quantities: carbon l_1, and cofactor A, l_6. There are four type B pools (Figure 10.8) that correspond to

l_2, high-energy conservation pool $(2C_6 + 3C_6P + 4C_6P_2 + 2C_3P_1 + 2C_3P_2 + C_3P + AP_3)$;

l_3, conservation of elemental P $(C_6P + 2C_6P_2 + C_3P_1 + 2C_3P_2 + C_3P + AP_3 + P)$;

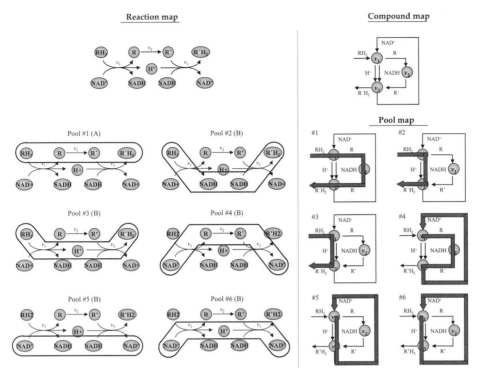

Figure 10.9: Schematic representation of the pool formation in multiple redox coupled reaction. Prepared by Iman Famili.

l_4, low-energy conservation pool $(2C_6 + C_6P + C_3P + 2C_3 + AP_2)$; and

l_5, potential to incorporate the standalone moiety $P(C_3P_2 + C_3P + C_3 + P)$.

l_2 and l_4 form energy conjugates to each other representing the high- and low-energy occupancy in the system. l_5 corresponds to l_3 in example V. The incorporation and exchange of phosphate thus results in four different convex basis vectors, each representing different aspects of the complex role of phosphate in energy metabolism (e.g., high-energy phosphate and standalone inorganic phosphate, as shown here). Note that the phosphate in AP_2 does not appear in the conservation of the elemental P, l_3, since AP_2 interacts as a whole moiety and is never reduced to other chemical moieties in this reaction network. In addition, the pool maps readily illustrate the reaction contributions to the conservation of biochemical moieties, such as shown by the pool map of l_5 that involves only v_4, v_5, and v_6.

Simplified TCA cycle

The TCA cycle is a circular pathway that converts a two-carbon unit into carbon dioxide and redox potential in terms of NADH. The convex basis for the left null space of the stoichiometric matrix for the simplified TCA network has five conservation moieties (Figure 10.8VII). They represent the conservation of

l_1, exchanging carbon group ($2H_2C_2 + 2H_2C_6 + HC_5 + C$);

l_2, recycled four-carbon moiety which "carries" the two carbon group that is oxidized ($C_4 + H_2C_6 + HC_5$);

l_3, hydrogen group that contains the redox inventory in the system ($2H_2C_2 + 2H_2C_6 + HC_5 + NH$);

l_4, redox vacancy ($C + N$); and

l_5, total cofactor pool ($N + NH$).

Note that l_3 can be partitioned into redox on carbon (i.e., $2H_2C_2 + 2H_2C_6 + HC_5$) and on the cofactor (NH). This partitioning then leads to a redox charge definition by taking the ratio of NH to ($N + NH$) forming a $NADH$ redox charge that is analogous to the energy charge.

10.6 Summary

➤ The left null space of **S** contains dynamic invariants.

➤ Integration of the time derivatives leads to a bounded affine space. All the concentration states, dynamic and steady, lie in this space of concentrations.

➤ Since all the concentrations are positive, a convex basis for this space is biochemically meaningful. A nonnegative convex basis for the left null space can be found.

➤ As with extreme pathways, three basic types of convex basis vectors can be defined for the concentration space.

➤ A suitable reference state can be defined for the affine concentration space to make it parallel to the left null space (and have the same basis) and orthogonal to the column space.

➤ The metabolic pools can be displayed on the compound map, leading to a representation that is similar to pathways on a flux map.

10.7 Further Reading

Clarke, B.L., "Stoichiometric network analysis," *Cell Biophysics*, **12**:237–253 (1988).

Clarke, B.L. "Stability of complex reaction networks." in ADVANCES IN CHEMICAL PHYSICS (I. Prigogine and S.A. Rice, Eds.), pp. 1–215, Wiley, New York (1980).

Famili I., and Palsson B.O., "The convex basis of the left null space of the stoichiometric matrix leads to the definition of metabolically meaningful pools," *Biophysical Journal*, **85**:16–26 (2003).

Schuster S., and Heinrich R., "Minimization of intermediate concentrations as a suggested optimality principle for biochemical networks. I. Theoretical analysis," Journal of matthematical Biology, **29**:425–442, (1991).

Schuster S., and Hilgetag C., "On elementary flux modes in biochemical reaction systems at steady-state," *Journal of Biological Systems,* **2**:165–182 (1994).

Schuster S., and Höfer T., "Determining all extreme semi-positive conservation relations in chemical reaction networks: A test criterion for conservativity," *Journal of the Chemical Society, Faraday Transactions*, **87**:2561–2566 (1991).

CHAPTER 11

The Row and Column Spaces of **S**

The column and row spaces of the stoichiometric matrix contain the concentration time derivatives and the thermodynamic driving forces, respectively. The basic content of these spaces is described in this chapter. A full analysis of these spaces requires knowledge of kinetics. Since we do not have comprehensive information about *in vivo* values of kinetic constants, the treatment of these spaces at the genome scale is necessarily limited.

11.1 The Column Space

The reaction vectors form the basis for the column space
The column space contains the derivatives of the concentrations of the compounds contained in a network. It is spanned by the reaction vectors ($\mathbf{s}_i$) as

$$\frac{d\mathbf{x}}{dt} = \mathbf{s}_1 v_1 + \mathbf{s}_2 v_2 + \cdots + \mathbf{s}_n v_n \tag{11.1}$$

as weighted by the fluxes through each of the reactions at any given instant in time. We note therefore that the different flux levels or the changes in the flux levels determine the location of the vector of derivatives in the column space.

Reactions with high flux levels and those changing much in time will thus generate large motion along the corresponding reaction vector. The reaction vectors can be organized by the expected size and responsiveness of the reactions. Although network dynamics are outside the scope of this book, we note that reactions that are fast and quickly come to some sort of a quasi-steady-state effectively reduce the column space's dimension on slower time scales. Time scale separation results, and for every dimension that the column space is reduced in this fashion, an effective additional dimension in the left null space is created.

The values of the individual fluxes are constrained, i.e., $v_i \leq v_{i,\max}$. Thus, the weighting factors in the sum of the reaction vectors in equation 11.1 are constrained and therefore the values of the time derivatives are constrained. Hence, the column space is a closed space.

Simple examples

Let us consider the reaction

$$2H_2O_2 \rightarrow O_2 + 2H_2O \tag{11.2}$$

we know that the left null space will be spanned by the elemental matrix

$$
\begin{array}{cccc}
 & H_2O_2 & O_2 & H_2O \\
O & 2 & 2 & 1 \\
H & 2 & 0 & 2
\end{array}
$$

We can use these vectors and the reaction vector $\mathbf{s}_1^T = (-2, 1, 2)$ to transform the concentration vector:

$$
\begin{pmatrix} -2 & 1 & 2 \\ 2 & 2 & 1 \\ 2 & 0 & 2 \end{pmatrix} \frac{d}{dt} \begin{pmatrix} H_2O_2 \\ O_2 \\ H_2O \end{pmatrix} = \frac{d}{dt} \begin{pmatrix} -2H_2O_2 + O_2 + 2H_2O \\ 2H_2O_2 + 2O_2 + H_2O \\ 2H_2O_2 + 2H_2O \end{pmatrix}
$$

$$
= \frac{d}{dt} \begin{pmatrix} R \\ O \\ H \end{pmatrix} = \begin{pmatrix} 9 \\ 0 \\ 0 \end{pmatrix} v_1
$$

where R is a group of concentrations changing over time.

The transforming matrix

$$
\begin{pmatrix} \mathbf{s}_1 \\ \mathbf{l}_1 \\ \mathbf{l}_2 \end{pmatrix} = \begin{pmatrix} -2 & 1 & 2 \\ 2 & 2 & 1 \\ 2 & 0 & 2 \end{pmatrix} \tag{11.3}
$$

represents the reaction vector and a convex basis for the left null space. We note that since

$$\langle \mathbf{s}_1 \cdot \mathbf{l}_1 \rangle = 0 \quad \langle \mathbf{s}_1 \cdot \mathbf{l}_2 \rangle = 0 \quad \langle \mathbf{l}_1 \cdot \mathbf{l}_2 \rangle = 4 \tag{11.4}$$

the reaction vector is orthogonal to the basis vectors of the left null space. However, the two chosen basis vectors for the left null space are not orthogonal.

The progress of this reaction can be graphically depicted, (Figure 11.1). The time derivative of R is nonzero while the time derivatives of the oxygen

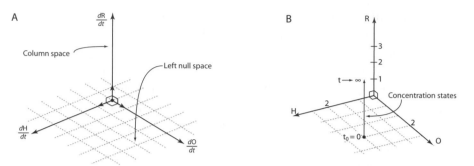

Figure 11.1: A graphical representation of the column and left null spaces, and the concentration states for the reaction $2H_2O_2 \rightarrow O_2 + 2H_2O$. (A) shows the time derivatives and (B) shows the concentration states, where an initial concentration vector of $(1, 0, 0)$ was used for the integration. Prepared by Iman Famili.

and hydrogen are zero. All the concentration states fall on a finite line whose coordinate is $\xi_1 \in [0, 1]$.

Reconsider the reaction

$$
X_1 + X_2 \overset{v_1}{\underset{v_2}{\rightleftharpoons}} X_3 \tag{11.5}
$$

In Chapter 8 we showed that

$$
\frac{d}{dt}\begin{pmatrix} X_1 \\ X_2 \\ X_3 \end{pmatrix} = \begin{pmatrix} -1 \\ -1 \\ 1 \end{pmatrix}(v_1 - v_2) + \begin{pmatrix} 0 \\ 0 \\ 0 \end{pmatrix}(v_1 + v_2) \tag{11.6}
$$

A transformation matrix

$$
\begin{pmatrix} -1 & -1 & 1 \\ 1 & 0 & 1 \\ 0 & 1 & 1 \end{pmatrix} \tag{11.7}
$$

leads to

$$
\frac{d}{dt}\begin{pmatrix} -x_1 - x_2 + x_3 \\ x_1 + x_3 \\ x_2 + x_3 \end{pmatrix} = \frac{d}{dt}\begin{pmatrix} R \\ l_1 \\ l_2 \end{pmatrix} = \begin{pmatrix} 4 \\ 0 \\ 0 \end{pmatrix}(v_1 - v_2)\begin{pmatrix} 0 \\ 0 \\ 0 \end{pmatrix}(v_1 + v_2)
$$

This fully decomposed system shows the participation of the flux vector and the representation of the conservation quantities and the dynamic motion of the reaction group R.

Next, consider two simultaneous reactions

$$
ADP \rightleftharpoons ATP \quad \text{and} \quad 2ADP \rightleftharpoons ATP + AMP \tag{11.8}
$$

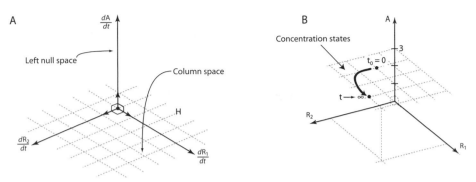

Figure 11.2: A graphical representation of the column and left null spaces, and the concentration states for the reaction system ATP $\rightleftharpoons$ ADP and ATP $+$ AMP $\rightleftharpoons$ 2ADP. (A) shows the time derivatives and (B) shows the concentration states, where an initial concentration vector of (1,1,1) was used for the integration. Prepared by Iman Famili.

that represent the use and production of ATP and the redistribution of the phosphate group among ATP, ADP, and AMP. For the moment, we leave out phosphate in the first reaction to make this a three compound system that can be shown graphically. The left null space is spanned by $(1, 1, 1)$, which represents conservation of the adenosine moiety.

We can use the two reaction vectors, and the basis of the left null space to transform the concentration vector:

$$\begin{pmatrix} 1 & -1 & 0 \\ 1 & -2 & 1 \\ 1 & 1 & 1 \end{pmatrix} \frac{d}{dt} \begin{pmatrix} \text{ATP} \\ \text{ADP} \\ \text{AMP} \end{pmatrix} = \begin{pmatrix} \text{ATP} - \text{ADP} \\ \text{ATP} - 2\text{ADP} + \text{AMP} \\ \text{ATP} + \text{ADP} + \text{AMP} \end{pmatrix} = \frac{d}{dt} \begin{pmatrix} \text{R}_1 \\ \text{R}_2 \\ \text{A} \end{pmatrix}$$

The progress of this reaction system can be graphically depicted in three dimensions (Figure 11.2). The time derivatives of R_1 and R_2 are nonzero, while the time derivatives of the adenosine moiety are zero. The concentration space is a finite plane whose coordinates can be represented by $\xi_1 \in [0, 1]$ and $\xi_2 \in [0, 1]$. Note that if one of the reactions is "fast" compared to the other, the path taken in the concentration space is L-shaped, as shown in Figure 11.2B.

11.2 The Row Space

A basis for the row space

The individual reaction fluxes form an orthogonal basis for the row space. This basis is a good choice when considering the stoichiometric mapping itself. Each reaction has a natural thermodynamic basis vector once the kinetics are included. As discussed in Chapter 9, each flux has a maximum value, confining all flux vectors to a rectangle in the positive orthant of the flux space. The null space lies within these confines and its orthogonal complement is the row space.

Constraints on the flux values

The magnitude of the individual fluxes is constrained. These constraints are derived from the limitation on the concentrations of the reactants and upper limits on the numerical values of the kinetic constants. The turnover rate of an enzyme complex (X)

$$v = k_1 x \le k_1(x + e) = k_1 e_{\text{total}} \tag{11.9}$$

where the total amount of enzyme (e_{total}) present is limited to $x + e$.

For a bilinear association reaction of a substrate to an enzyme, the rate is

$$v = k_b x_i e \le k_b a_i e_{\text{total}} \tag{11.10}$$

where a_i is the size of the most limiting conservation pool of which x_i is a member. Limiting the turnover rates would be a more effective way to control fluxes through enzymatic pathways since only one variable needs to be regulated: the total amount of enzyme. The release step of the product from the enzyme often turns out to be the rate limiting step in enzyme catalysis [114].

Thermodynamic driving forces

If the fluxes are imbalanced, there will be a net generation or elimination of compounds in the network. If we denote a row in **S** by $\mathbf{r}_i$, then the corresponding dynamic mass balance is

$$\frac{dx_i}{dt} = \langle \mathbf{r}_i \cdot \mathbf{v} \rangle \tag{11.11}$$

thus, the time derivative is the inner product of that row vector and the flux vector:

$$\langle \mathbf{r}_i \cdot \mathbf{v} \rangle = \| \mathbf{r}_i \| \| \mathbf{v} \| \cos(\theta_i) \tag{11.12}$$

where θ_i is the angle between the two vectors. If this inner product is zero, then the flux vector is orthogonal to the row vector. If not, this inner product sets the magnitude of the time derivative. Geometrically, the magnitude of this inner product may be viewed as a projection of the flux vector on the row vector. Since the magnitude of the row vector is fixed and the individual fluxes are bounded, this inner product is also bounded.

11.3   Summary

➤ The column space is naturally spanned by the reaction vectors.

➤ The row space can be represented by an orthogonal basis formed by the individual fluxes with values only in the positive orthant.

➤ The magnitude of the individual fluxes is limited by kinetics and caps on concentration values.

➤ This limitation also limits the possible values of the time derivatives and thus the column space.

➤ The column and row spaces are closed.

11.4   Further Reading

Wei, J., and Prater, C.D., "The structure and analysis of complex reaction systems," *Advances in Catalysis*, **13**:203–392 (1960).

PART THREE

Capabilities of Reconstructed Networks

The functional states of reconstructed networks are directly related to cellular phenotypes. With reconstructed networks represented in the form of **S**, we can use mathematics to compute their candidate functional states of reconstructed networks. If one adopts the informatics point of view of **S** and its annotated information as biochemically, genetically, and genomically (BIGG) structured database, then these *in silico* methods are viewed as *query* tools. Whether viewed from an informatic or mathematical standpoint, the result of applying *in silico* analysis methods is the study of *network properties*, sometimes called *emergent properties*. These properties represent functionalities of the whole network and are hard to decipher from a list of its individual components. In some sense, these properties are a reflection of the hierarchical nature of living systems. A variety of methods have been developed to examine the properties of genome-scale networks. The third part of this text summarizes the *in silico* methods that have been developed and deployed to date. The development and application of such methods is the focus of a growing number of researchers worldwide, and we can thus anticipate that there will be much progress in this field over the coming years.

CHAPTER 12

Dual Causality

The stoichiometric matrix and its associated information fundamentally represents a biochemically, genetically, and genomically structured database. It can be used to analyze network properties, and to relate the components of a network and its genetic bases to network or phenotypic functions. In developing biologically meaningful *in silico* analysis procedures, fundamental characteristics of biology need to be explicitly recognized. Unlike the physicochemical sciences, biology is subject to *dual causality* or *dual causation* [136]. Biology is governed not only by the natural laws but also by genetic programs. Thus, while biological functions obey the natural laws, their functions are not predictable by the natural laws alone. Biological systems function and evolve under the confines of the natural laws according to basic biological principles, such as the generation of diversity and natural selection. The natural laws can be described based on physicochemical principles and used to define the constraints under which organisms must operate. How organisms operate within these constraints is a function of their evolutionary history and survival. Survival and its relationship to cellular functions can perhaps be readily understood for simple, single cellular organisms.

12.1 Causation in Physics and Biology

Physics

Classically, "cause and effect" is established by formulating mathematical descriptions of conceptual models of fundamental physical phenomena. One example is that of molecular diffusion; see Figure 12.1. The fundamental process underlying diffusion is the random walk process that a collection of molecules undergoes. The statistical properties of the random walk

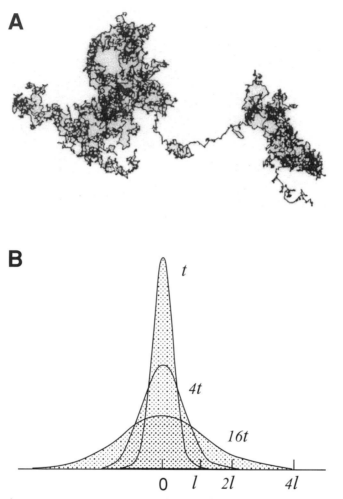

Figure 12.1: Random walk and diffusion. (A) The simulated trajectory of a single molecule. (B) The probability distribution for the molecule's location as a function of time when it was located at $l = 0$ at $t = 0$. The width of the distribution, l, increases with the square root of time. Modified from [156].

process can be quantitatively assessed, and its macroscopic consequences are described with Fick's law. Fick's law is described by a simple equation that is used as the basis to describe mass transfer processes. The established causality is the basis for computations that reliably predict the consequences of mass transfer processes. Engineering design can be based on such predictions. The Boltzmann and Nernst equations provide other specific cases of causality in physics, and there are many more examples. Thus, in physics and engineering "there is nothing more practical than a good theory." Cause and effect for physical phenomena are thus often well established and can be described mathematically. Mathematical

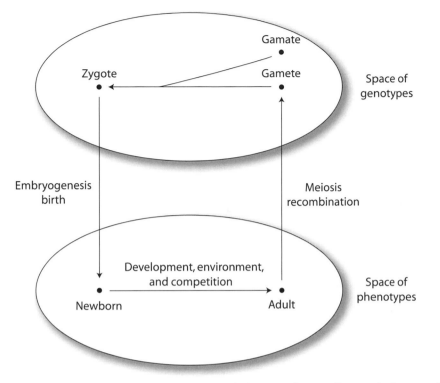

Figure 12.2: Mapping between genotype and phenotype in sexually reproducing organisms. This figure illustrated the iterative scrambling of genes and the selection of the resulting phenotype. Modified from [210].

descriptions are often in the form of equations and inequalities. Discussion of the character of physical law is available [58].

Biology

Causation in biology is much different than in physics. Biological causation fundamentally originates from replication or reproduction that produces nonidentical offsprings. In particular, sexual reproduction is a very efficient process to constantly recombine chromosomes to produce nonidentical individuals; see Figure 12.2. This process leads to diversity, a biopopulation of nonidentical individuals. Natural selection then determines survival, which, over time, leads to evolution of the species.

Thus, living systems are time variant, meaning they evolve and change over time. In contrast, physical phenomena are time-invariant; e.g., oxygen diffuses the same way in water under given circumstances, and the unit charge on the electron does not change. The omnipresent selection process

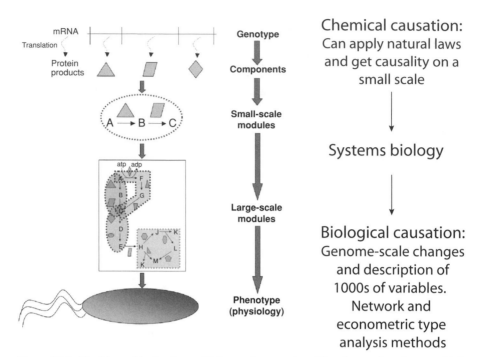

Figure 12.3: The hierarchical nature of living systems and multilevel causation. Systems biology tries to provide structure to the hierarchical relationship between molecular and physiological events. Prepared by Timothy Allen.

in biology gives the appearance of "sense of purpose" that is fundamentally survival. The outcome of the selection process is in part stochastic and is influenced by environmental variables. Such attributes are not given to fundamental physical phenomena.

Hierarchy

Thus, we conceptualize causation in physics and biology differently. However, both are relevant to systems biology and they represent opposite ends of a hierarchical process (Figure 12.3). As discussed by F. Jacob [102] constraints are upwardly applicable in the hierarchy. In other words, processes like diffusion, that constrain intracellular processes [242], thus constrain all, but do not specify, higher level functions. Similarly, Figure 2.7 shows that early events in evolution tend to constrain subsequent evolutionary events. Hierarchical organization, a fundamental biological principle, makes it difficult to interpret, let alone predict, the outcome of selection processes at higher levels of biological complexity (recall Section 2.4). However, a constraint-based analysis provides a framework within which these basic considerations can be accommodated.

12.2 Model Building in Biology

Because of dual causality, mathematical model building in biology at the network and genome scale will need to differ from that practiced in the physicochemical sciences. In these fields, one starts with basic physicochemical principles, such as thermodynamics, chemical potential, the diffusion equation, mass conservation, or the Nernst and Boltzmann equations. These equations are based on well-developed fundamental physical theories, and they typically contain a large number of parameters, most of which can be individually measured under defined conditions. These equations then form the basis for computer models and simulation.

Limitations of theory-based modeling approaches

In spite of the impressive bioinformatic databases that are currently available, we cannot obtain all of the information needed to build a detailed computer model of a whole cell based on physicochemical principles. Traditional theory-based models of large-scale cellular processes are faced with fundamental challenges:

First, the intracellular chemical environment is complex (e.g., see [76, 77]) and hard to define in terms needed for the formulation of equations that describe the physics of the intracellular milieu.

Second, assuming that we had all the governing equations defined, we would have to find numerical values for all the parameters that appear in these equations. These values would have to be accurate for intracellular conditions.

Third, even if we could overcome the first two challenges, we would face the fact that evolution changes the numerical values of kinetic constants over time. In addition, in a biopopulation, we could have a perfect *in silico* model for one individual organism, but it would not apply to other individuals in the biopopulation due to genetic and epigenetic differences between individuals. Such time dependency and diversity of parameter values are key distinguishing features between biological and physicochemical systems.

Constraining behaviors

The third limitation results from the dual causality that needs to be accounted for in realistic models of biological processes. An approach to the *in silico* analysis of cellular functions can be formulated that is based on the fact that cells are subject to governing constraints that limit their possible behaviors. Imposing these constraints, it can be determined what

Table 12.1: Physicochemical factors constraining the functions of biochemical reaction networks. Adapted from [152].

Factor	Type of constraint
Physicochemical constraints Osmotic pressure, electroneutrality, solvent capacity, molecular diffusion	Hard nonadjustable constraints
Connectivity Systemic stoichiometry Causal relationships	Hard nonadjustable constraints
Capacity Maximum/minimum flux	Nonadjustable maximum based on maximum association rates Adjustable by transcriptional regulation
Rates Mass action, enzyme kinetics, regulation	Highly adjustable by an evolutionary process

functional states can and cannot be achieved by a reconstructed network. Imposing a series of successive constraints can limit allowable cellular behavior, but will never predict it precisely.

The imposition of constraints leads to the formulation of *solution spaces* rather than the computation of a *single solution*, the hallmark of theory-based models. Cellular behaviors (i.e., functional states) within the defined solution space can be attained. Each allowable behavior basically represents a different phenotype based on the component list, the biochemical properties of the components, their interconnectivity, and the imposed constraints. The constraint-based approach leads to *in silico* analysis procedures that are helpful in analyzing, interpreting, and occasionally predicting the genotype–phenotype relationship.

Cells are subject to a variety of constraints. There are both *nonadjustable* (i.e., invariant or hard) and *adjustable* constraints (see Table 12.1). The former can be used to bracket the range of possible behavior, and the latter can be used to further limit allowable behavior, but these constraints can adjust through an evolutionary process or through changing environmental conditions. In addition, the adjustable constraints may vary in a biopopulation from one individual to another.

Successive imposition of constraints

Constraints can be applied to the analysis of reconstructed networks to narrow attainable behaviors, and can be applied in a successive fashion,

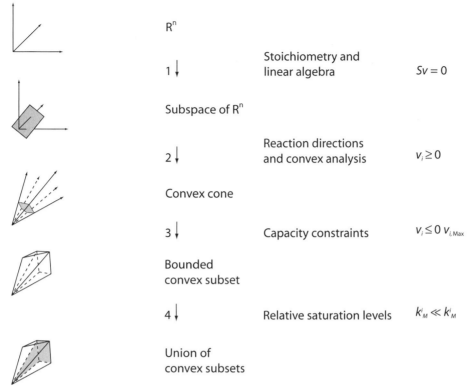

R^n		
$1\downarrow$	Stoichiometry and linear algebra	$Sv = 0$
Subspace of R^n		
$2\downarrow$	Reaction directions and convex analysis	$v_i \geq 0$
Convex cone		
$3\downarrow$	Capacity constraints	$v_i \leq 0\, v_{i,Max}$
Bounded convex subset		
$4\downarrow$	Relative saturation levels	$k_M^i \ll k_M^i$
Union of convex subsets		

Figure 12.4: Narrowing down alternatives. The successive application of constraints to a reaction network narrows down the attainable functional states. Redrawn from [152].

as illustrated in Figure 12.4. The first icon in Figure 12.4 shows a space where the axes represent fluxes through reactions in a network. Not all the points in this space are attainable due to the interrelatedness of the fluxes. The flux balances limit (through an equation $\mathbf{Sv = 0}$) the steady-state fluxes to a subspace that is a hyperplane (step 1). If the reactions are defined so that all the fluxes are positive, this plane is converted into a semifinite space, called a cone (step 2). The edges of this cone become a set of unique, systemically defined extreme pathways, as detailed in Chapter 9. All the points inside the cone can be represented as nonnegative combinations of these extreme pathways. Since there are capacity constraints on the individual reactions in the extreme pathways, the length of each edge is limited. This closes the cone (step 3) and forms a closed solution space in which all allowable network states lie. The properties of this space can then be studied using the methods described in the subsequent three chapters. Further segmentation of this space can be achieved if additional constraints, such as kinetic constants or thermodynamic quantities, are available (step 4).

Table 12.2: Generations of genome-scale models.

Model generation (and their use)	Type of data used
First: (what states are possible)	Genomic and legacy
Second: (what states are chosen)	Transcriptomic and genome location
Third: (how chosen states are achieved)	Time-dependent proteomic and metabolomic

Developing genome-scale models

The successive application of constraints lends itself to a step-wise development of increasingly refined data-driven *in silico* models. These models broaden in scope with the establishment and imposition of additional constraints. Constraint-based models can address the questions relating to allowable functions of a given network, which of these functions the cell actually chooses, and how such choices are made (Table 12.2).

The *first generation* of constraint-based models for microbial metabolism have appeared [110, 175, 183]. The "omics" data type on which they are based is *genomic*. Literature (legacy) data are used as well as the formulation of hard physicochemical constraints, such as mass, energy and redox balance, thermodynamic, and maximal reaction rates. These constraints collectively define all the possible functionalities of a reconstructed network, which are mathematically confined to a solution space. The properties of this space can be studied by the methods described in the following chapters.

The *second generation* of constraint-based models include the imposition of transcriptional regulatory networks, leading to the shrinking of the allowable states of metabolic networks. In response to environmental queues and built-in regulation, the solution space is shrunk [35, 36] to contain network functions that the cell has chosen through an evolutionary process; recall Figure 2.5. The choices that a cell makes can then be identified and analyzed.

The *third generation* of constraint-based models are likely to account for the abundance or concentration of the cellular components. Various "omic" data types can now be obtained in a time-resolved fashion. Such data will help clarify just how the cell implements the choices it has made and how it evolves to find new choices. This approach is likely

Table 12.3: The engineering design procedure contrasted with the constraint-based modeling approach. Inspired by Edwin Lightfoot.

Engineering design	Constraint-based modeling
Statement of goal	Survival criteria (fitness)
Statement of constraints (time, money, shape, materials, etc.)	Identification of constraints (physicochemical, topological, environmental, regulatory)
Define design envelope	Formulate solution space
Choose design variables to produce best design	Evolve toward best fitness (by adjusting regulation and kinetic parameters)

to lead to the definition of the rate constants of the network as a whole rather than constants for the individual underlying biochemical events.

First generation constraint-based models of metabolism in several microbes have been developed [110, 175, 183]. These models have generated several valuable results. For instance, the initial models predicted about 60% of knock-out growth phenotypes correctly in *Helicobacter pylori* [192], 86% in *Escherichia coli* [51, 185], and about 85% in *Saccharomyces cerevisiae* [54, 66]. These were simple, qualitative, growth/no growth predictions. Using the genome-scale *in silico* model of *E. coli*, experimentally testable predictions were formulated describing the quantitative relationship between the primary carbon source uptake rate, oxygen uptake rate, and cellular growth rate. Experimental data for *E. coli* were consistent with optimal use of the network to maximize growth under the conditions examined [51, 63]. The outcomes of adaptive evolutionary processes have also been analyzed using first generation models, by placing cells under growth rate selection over a long period of time [64, 99].

Through the reconstruction of transcriptional regulatory networks (Chapter 4), we can now begin to impose condition-dependent restraints on reconstructed metabolic networks [38, 33] and formulate the second generation models. These transcriptional regulatory networks represent a mechanism by which the cell makes choices in response to changes in the internal or external states of the cell. Mathematically, they lead to a shrinking of the solution space so that it contains only the solutions or phenotypes that the cell chooses to utilize under a particular condition.

A limited analogy to the engineering design process

The constraint-based modeling procedure detailed above has some striking similarities with the engineering design process; see Table 12.3. An engineering design begins with the statement of a goal, i.e., achieve the separation of all the protein in *E. coli*, or build a bridge across the Golden Gate. Then one states the constraints under which the design is being produced. Time and money are always major constraints, but so are shape requirements and the availability of materials. After the constraints are specified, the design, or free variables and their allowable numerical values are identified, resulting in the definition of the design envelope. The design variables are then varied to produce the best design based on some stated objective, such as cost and functionality.

The constraint-based modeling procedure consists of the same fundamental steps. The analogy between the two is, of course, limited. There are at least two major differences. Biology, by necessity, employs a one-step look-ahead procedure, that is probabilistically based through the generation of alternatives. Then selection follows. The engineer, by contrast, can employ a multistep look-ahead strategy. The goal for engineering design is typically well defined as opposed to the survival objectives of cells, and different survival strategies may succeed. Furthermore, organisms must deal with a time-varying environment. Nevertheless, this analogy points to the adjustable constraints, such as changing kinetic and regulatory parameters, as the major biological design variables. It should be noted that engineering designs evolve over time as well.

Redundancy, multifunctionality, and noncausality

Biological networks have several fundamental properties that need to be considered when interpreting large-scale data sets and building models to describe their functions.

Redundancy: Biochemical reaction systems have redundancy built into them at many levels. Often, individual steps in a network can be carried out in more than one way. Isozymes represent different enzymes that carry out the same reaction. Similarly, some codons can be translated by more than one tRNA. There are also network-level redundancies. The overall function of a network to support a phenotype can be achieved in more than one way. Thus, there are multiple equivalent outcomes from the same biological selection process. The mathematical aspect of this feature, *alternate optimal solutions*, is detailed in Chapter 16. Biologically, these can be called *silent phenotypes*.

Multifunctionality: There are components in biochemical networks that can carry out more than one function. Examples of this feature include enzymes that can catalyze many related chemical reactions. Similarly, some tRNA molecules can translate more than one codon. At the network level, there could be global network states that would give similar phenotypes even if the environments are different. This feature would be called a *generalist phenotype*. The notion of a high-flux back-bone in metabolism [2] comprises a set of reactions that lead to optimal growth on different substrates. A high-flux backbone is an example of a large correlated set of reactions that function together in optimal solutions [159].

Noncausality: Due to the hierarchical organization of organisms, changes on one level may not percolate up to functions at a higher level of organization, and would thus be noncausal. A well-known example of non-causality is *hitchhiker mutations* that co-select with a causal mutation located nearby on the genome. It is not known if the same is true for individual variables changing in high-throughput data sets, such as expression profiling. In the field of signal transducton, many are interested in knowing "who-talks-to-whom," meaning that one wants to know all possible chemical interactions between two components. Protein–protein interaction maps provide one example. However, the biologically–meaningful question is "who-listens-to-whom," since we are only interested in knowing if chemical interactions are a part of a higher order biological function. Thus, there can be many noncausal (biologically), but detectable, chemical interactions among macromolecules.

These three attributes are important considerations in studying the hierarchical nature of biological systems. Multiscale, multiparameter analysis methods will be needed to study this hierarchical organization. They will need to be able to deal with nonregular patterns, which will be a deviation from classical methods such as Fourier analysis that looks for repeated regular patterns. All of these features have appeared through the evolutionary process, which abides by a series of constraints.

12.3 Models Can Drive Discovery

A number of genome-scale networks have been reconstructed based on available data; see Table 3.6. At present, there is no organism for which a complete data set exists. Therefore, the *in silico* analyses of these reconstructions will lead to failed predictions and enable data-driven model updates. In this way, the reconstruction as a chemically and genetically defined database offers a framework to identify gaps in our knowledge about an organism.

Failure modes
From a discovery standpoint, the primary results from genome-scale analysis using constraint-based analysis methods are the failure modes. Detailed

analysis of failure modes leads to iterative model building, Figure 12.5. For instance, the study of failure modes from the analysis of knock-out strain phenotypes can be seen in [51, 67], updated sequence annotations based on *gap analysis* in [185], and phenotype prediction failures in [47, 33]. Such analysis leads to well-defined questions that are addressed during the next iteration of the model building process. The analysis of failure modes and the design of experiments to resolve them have not yet been algorithmatized.

The iterative model building process

The process of building mathematical models and running computer simulations of complex biological processes is iterative. Reconstructed "*in silico* organisms" are computer representations of their *in vivo* counterparts. They are based on the curation of available data to form what is basically a biochemically and genetically structured database. These *in silico* models will have some analytical, interpretive, and predictive capabilities. However, due to incomplete knowledge of constraints and incomplete or erroneous annotation, genome-scale *in silico* models will only be able to represent some functions of the organism correctly.

One must therefore learn to appreciate the value of failure in prediction. The main difference between the *in silico* and *in vivo* organism is that the *in silico* version is incomplete and missing some features. Therefore, we must set out to formulate experimentally testable hypotheses based on the *in silico* analysis, perform the experiments, and update the models (Figure 12.5). Interestingly, this iterative process in building *in silico* organisms is likely to have two loops. One is the classical experimental loop (on the right in Figure 12.5), and the other is *in silico* (on the left in Figure 12.5). Many corrections and adjustments to these models are likely to originate from analyzing and searching the ever-growing bioinformatic databases.

In silico models as hypotheses

The scientific process has traditionally been based on the statement of hypotheses. Hypotheses represent the result of an investigator's evaluation of the available data and past knowledge in the field. Well-defined experiments are then constructed in an attempt to invalidate the hypothesis.

In the era of high-throughput biology, it is now possible to generate volumes of data, much more than needed to address narrow hypotheses. This capability has led to the so-called *discovery-based approach* in biological and medical research. In this approach, the investigator obtains large amounts of data about a process or phenomenon, without an a priori statement of hypothesis. The data are subsequently analyzed, to look for

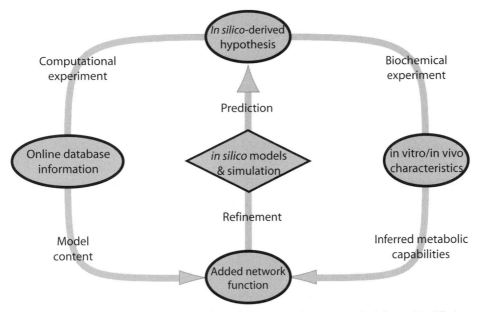

Figure 12.5: The iterative *in silico* model-building procedure as a dual loop. Modified from [152].

any interesting correlations, information, or content from which scientific "knowledge" can be derived.

As a result of high-throughput biology, a clash has emerged in the approach to the way biological research is conducted. Rigorous thinking in science would dictate that discovery-driven research is somehow a "second-class citizen" in the scientific endeavor. However, it has become clear that high-throughput biology can indeed be hypothesis driven. The hypotheses in the era of systems biology may not be the simple, crisp statements as have been seen in the past, but instead may be represented by complex *in silico* models.

It is difficult for an investigator to comprehend the vast high-throughput data sets that are being generated. The enormity of these data sets, often including hundreds of thousands or potentially millions of data points, is such that the human mind cannot keep track of all the variables involved. An *in silico* model, or a mathematical representation of the data, would serve as its most compact description. Like scientific hypotheses, these *in silico* models will include underlying mechanisms rather than being correlative statistical models. Such models represent highly structured hypotheses and the most compact representation of the data at hand.

In silico models, therefore, like classical hypotheses, are meant to be disproved. An interesting difference between a classical hypothesis in science and an *in silico* model as a hypothesis in systems biology is that only parts of the model are disproved, but not necessarily the model in its entirety. Thus, through a systematic procedure, pieces of a model are invalidated, and as the model is built and its capability of representing data improves, the more accurate representation of an organism it becomes.

Unlike physics and chemistry, it is often difficult to state crisp hypotheses about complex biological processes. This difficulty arises from the enormous complexity of any biological process and the difficulty breaking it down into small pieces. If the pieces are small enough, the hypotheses look no different than those in chemistry. For instance, hypothesizing whether a binding factor regulates two or more genes would simply come down to ascertaining a chemical binding affinity to a region of the genome. Conversely, *in silico* models as hypotheses allow us to address systems or integrated properties of a network and apply it at different levels in the hierarchy of biological systems (recall Figure 2.6).

Experimental designs to probe network functions

Experimental examination of network functions is often approached through *perturbation experiments*. Such experiments examine the function of a network by perturbing it in a defined fashion and then comparing the functions of the perturbed and unperturbed networks.

There are several types of perturbation experiments. We divide them here into two major categories:

Single variable perturbation: A single variable can be changed to examine the effects of individual components of a network. The two most frequently used variables are *genetic* and *environmental perturbations*. Genetic perturbations can be achieved by removing a gene from a genome to produce a knock-out strain, or by using methods to "silence" the gene product, such as through the use of small inhibitory RNA molecules. On a smaller scale, a base pair in the DNA sequence of a gene can be changed through site-directed mutagenesis to examine the effects of a point mutation in a controlled setting. Environmental variables are easy to control in a laboratory setting. A single medium component, such as carbon or nitrogen sources, can also be changed. Similarly, *heat shock* experiments can be performed by suddenly changing the temperature. A combination of gene and environmental perturbations (a *double perturbation experiment*) have been effectively used to examine causal relationships [101] and to iteratively build transcriptional regulatory networks [33].

Systemic perturbations: The network states resulting from the simultaneous charge in a number of variables can be compared. The state of networks in normal cells versus cells involved in a diseased state can be compared and used to guide discovery of *drug targets* that can then be used for the development of therapeutic compounds [95]. Similarly, familial background can be used to compare the responses of cells from a patient and those from the next of kin to reduce genetic differences to a minimum in the discovery of drug targets and for drug development [154]. Finally, adaptive evolution can be used to compare the changes in the functional states of networks as they improve in fitness through a rigorous selection process [61].

12.4 Constraints in Biology

The process of evolution is fundamental to the biological sciences. Organisms exist in particular environments, and as they replicate, they produce offsprings that are not genetically identical to the parent, thus generating a *biopopulation* of individuals that are each slightly different from one another. Over time, *natural selection* favors those individuals in the biopopulation that have more *fit* functions than other members of the biopopulation. To survive in a given environment, organisms must satisfy myriad constraints, which limits the range of available phenotypes. The better an organism can achieve a relatively fit function in a given environment (which includes the other members of the biopopulation), the more likely it is to survive.

All expressed phenotypes resulting from the selection process must satisfy the governing constraints. Therefore, clear identification and statement of constraints to define ranges of allowable phenotypic states provides a fundamental approach to understanding biological systems that is consistent with our understanding of the way in which organisms operate and evolve. Thus, in developing a mathematical framework within which to analyze organism functions, it becomes important to identify governing constraints. As pointed out in Section 2.4, several authors have discussed general constraints in biology [34, 41, 73, 102, 136]. Different types of constraints limit cellular functions. Here, we divide constraints into four categories [176]: fundamental physicochemical, spatial or topological, condition-dependent environmental, and regulatory, or self-imposed, constraints. We must be mindful of the fact that constraints alone cannot predict many basic biological features such as the genetic code [39].

Physicochemical constraints
Many physicochemical constraints govern cellular processes. These constraints are inviolable and thus represent hard constraints. Conservation of mass, energy, and momentum represent hard constraints. The interior of a cell is densely packed, forming an environment where the viscosity

may be on the order of 100–1000 times that of water. Diffusion rates inside a cell may be slow, especially for macromolecules. The confinement of a large number of molecules within a semipermeable membrane causes high osmolarity. Thus, cells require mechanisms for dealing with the osmotic pressure generated, such as sodium–potassium pumps to balance osmolarity or a rigid cell wall to physically withstand it. Intracellular reaction rates are determined by local concentrations inside cells. Reactions have maximal reaction rates (denoted with v_{max}) estimated to be about a million molecules per μm^3 per sec. Furthermore, biochemical reactions need to have a negative free energy change in order to proceed in the forward direction. These are some of the many basic physicochemical constraints that cells must satisfy.

Topobiological constraints

The crowding of molecules inside cells leads to *topobiological*, or three-dimensional, constraints; recall Figure 2.3. The linear dimension of the bacterial genome is on the order of 1000 times that of the length of the cell. DNA must therefore be tightly packed in the nuclear region in an accessible and functional configuration since DNA is only functional if it is accessible. Thus, at least two competing needs (to be tightly packed, yet accessible) constrain the physical arrangement of the bacterial genome. And there are likely to be additional constraints. As one further example, we note that the ratio between the total number of tRNA molecules and the number of ribosomes in a typical *E. coli* cell is approximately 10 [143]. With 43 different types of tRNA, there is less than one full set of tRNAs per ribosome. The genome, therefore, may have to be configured such that the location of rare codons is spatially close and translated by the same ribosome. Identification of these constraints and analysis of their consequences will be important for the understanding of the three-dimensional organization of cells.

Environmental constraints

Environmental constraints on cells, such as nutrient availability, pH, temperature, osmolarity, etc., are typically time and condition dependent. For example, *H. pylori*, a human gastric pathogen, lives in a relatively constant environment but is constrained by its low pH. It produces ammonia to sufficiently neutralize the pH in its immediate surroundings in order to stay alive. Conversely, the life cycle of *E. coli* is characterized by a series of sudden environmental changes. Outside of an animal it lives at ambient temperature and in the presence of ample oxygen. Then it experiences a heat shock when it enters the mouth of an animal, followed by an acid shock when it reaches the stomach. Following entry into the small intestine another pH shock is experienced, followed by a nutritionally rich anaerobic environment where it can grow rapidly in the presence of other bacterial species. Then finally it experiences a cold shock and ample oxygen with diminishing nutrients surrounding it. *E. coli* needs to be able to adjust its internal functional state to survive this series of environmental changes.

Knowing the environmental constraints is of fundamental importance for the quantitative analysis of microorganism functions; however, natural environments may be hard to precisely define. Conversely, in the laboratory, defined growth media can be used so that the environmental variables are precisely known. Laboratory experiments with undefined media composition are often of limited use for quantitative *in silico* modeling.

Regulatory constraints

These constraints are fundamentally different than the three types discussed above. They are *self-imposed* and are subject to evolutionary change, and can thus be time variant. For this reason, these constraints may be referred to as regulatory *restraints*, in contrast to the physicochemical constraints, the topological constraints, and time-dependent environmental constraints. Based on environmental conditions, regulatory constraints provide a mechanism to eliminate suboptimal phenotypic states (recall Figure 2.5) and confine cellular functions to behaviors of high fitness. Regulatory constraints are produced in a variety of ways, including the amount of gene products present and their activity; recall Figure 3.9.

Mathematical representation of constraints: balances and bounds

Following their definition, governing constraints need to be described mathematically. Then they can be used to perform *in silico* analyses. Mathematically, constraints are represented as either *balances* or *bounds*.

- A balance constraint is represented by an equality such as the conservation of mass. In a steady state, there is no accumulation or depletion of compounds; thus, the rate of production equals the rate of consumption for each compound in the network. This balance is represented mathematically as $\mathbf{S}\mathbf{v} = 0$, as detailed in Chapter 6. Similar balance equations can be formulated for other quantities, such as osmotic pressure, electroneutrality, and free energy around biochemical loops.

- A bound is represented by an inequality. Bounds are constraints that limit the numerical ranges of individual variables and parameters such as concentrations, fluxes, or kinetic constants. Upper and lower limits can be applied to individual fluxes ($v_{\min} \le v \le v_{\max}$). For elementary (and irreversible) reactions $v_{\min} = 0$, and $v_{\max}$ is less than approximately one million molecules per μm^3 per sec. Concentrations must

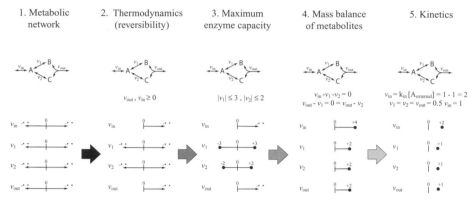

Figure 12.6: Narrowing down alternatives, a simple example. Taken from [34].

always be nonnegative; so $0 \leq x_i$. Upper bounds for concentrations can arise from solvent capacity constraints, represented as $x_i \leq x_{max}$. Kinetic constants are constrained to be positive and have an upper bound based on collision frequency ($0 \leq k \leq k_{max}$). Transmembrane potentials are limited to about 240–270 mV, above which lipid bilayers destabilize.

Illustrative example

The details of the successive imposition of constraints can be illustrated with a simple example. A small network with only two chemical reactions (flux $v_1 : A \rightarrow B$ and flux $v_2 : A \rightarrow C$) as well as two transport processes (metabolite A enters the cell with flux v_{in} and B and C exit together via flux v_{out} through a symporter) is depicted in Figure 12.6. The candidate functional states of this toy network can be defined through the imposition of successive constraints:

1. Initially following the reconstruction, there is nothing that constrains the flux values.
2. If the exchange fluxes are irreversible, then v_{in} and v_{out} become nonnegative variables (i.e., they cannot take on negative values that would restrict the numerical values that they can take).
3. If the capacities of the internal enzymes are known (i.e., 3 for v_1 and 2 for v_2 in Figure 12.6), the numerical range for their flux values becomes finite.
4. If the network is in a steady state, a flux balance is imposed further restricting the allowable ranges of the flux values. Note that the exchange fluxes are now confined to a closed range, and that this range corresponds to the closed null space (see Chapter 9) for this toy network.

5. If the system is completely characterized, and all numerical values for the physical parameters are known, then the allowable numerical ranges are reduced to a single point, and we have a unique solution.

12.5 Constraint-Based Analysis Methods

A variety of *in silico* procedures have been developed to study the properties of genome-scale networks [176]. These methods are detailed in the remaining chapters of this book. They are organized and summarized in Figure 12.7.

Reconstruction and imposition of constraints: At the heart of constraint-based procedures is network reconstruction and the use of constraints to define a solution space. These two core topics are covered in Part I and this chapter, respectively.

Properties of solution spaces: Once the solution space is formed, its properties can be studied. Chapter 9 described the definition of a conical set of basis vectors that can be used to generate the closed solution space. This set of basis vectors can be used to study various network properties (Chapter 13). The defined solution spaces can also be uniformly and randomly sampled to generate a set of representative solutions in the space. The properties of this set of solutions can then be studied (Chapter 14). These methods amount to an *unbiased* assessment of the defined solution space and its overall properties.

Finding particular network states: One may want to find particular solutions in the space that represent network states of interest. Thus, a *biased* search for these states is carried out. Computationally, such a search is carried out using constrained optimization. If the constraints used are linear equalities and inequalities, then the solution space is a polytope. If the solutions sought can be described by a linear *objective function*, then the optimization process is carried out using *linear programming* (LP). This procedure to find particular solutions is the popular *flux-balance analysis* (FBA). It is described in Chapter 15.

Parametric sensitivity analysis: FBA was the first method developed in a series of optimization-based approaches to study particular aspects of the solution spaces associated with biochemical reaction networks. Within this framework, a plethora of *in silico* analysis methods have been developed. These include the study of parametric sensitivities, the consequences of gene deletions, and changes in environmental conditions. These methods are detailed in Chapter 16.

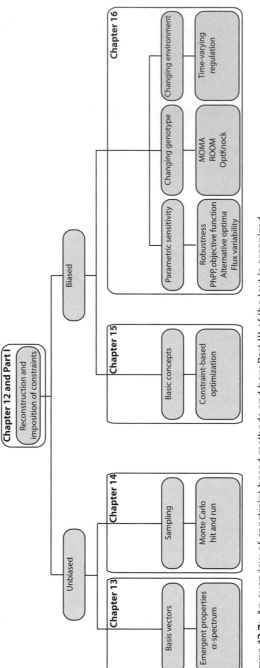

Figure 12.7: An overview of constraint-based methods and how Part III of the text is organized.

12.6 Summary

➤ Dual causation in biology requires us to accommodate the physico-chemical constraints under which cells operate, as well as fundamental biological properties, such as natural selection and generation of alternatives.

➤ Organisms have to abide by a series of constraints, including those arising from basic natural laws, spatial constraints, and the environment in which they operate. Many possible biological functions are achievable under these constraints, and organisms willfully impose constraints through various regulatory mechanisms to select useful functional states from all allowable states.

➤ A constraint-based approach emerges from these considerations that enables the simultaneous analysis of physicochemical factors and biological properties.

➤ Genome-scale reconstructions are mathematically represented and the governing constraints are imposed. This procedure leads to an *in silico* organism that contains all the known components of the real organism that it represents, and allows the simulation of allowable states given a set of governing constraints.

➤ The *in silico* model building process is iterative and will proceed in multiple steps.

➤ *In silico* models function as a structured way of integrating data and systematically building hypotheses.

➤ A wide range of constraint-based analysis methods have appeared and are being used to analyze various aspects of genome-scale models and the biological properties of the organisms that these models represent.

12.7 Further Reading

Covert, M.W., Famili, I., and Palsson, B.O., "Identifying constraints that govern cell behavior: A key to converting conceptual to computational models in biology?," *Biotechnology & Bioengineering*, **84**:763–772 (2003).

Hood, L., and Perlmutter, R.M., "The impact of systems approaches on biological problems in drug discovery," *Nature Biotechnology*, **22**:1215–1217 (2004).

Kauffman, K.J., Prakash, P., and Edwards, J.S., "Advances in flux balance analysis," *Current Opinion in Biotechnology*, **14**:491–496 (2003).

Lazebnik, Y., "Can a biologist fix a radio? – Or, what I learned while studying apoptosis," *Cancer Cell*, **2**:179–182.

Mayr, E., THIS IS BIOLOGY: THE SCIENCE OF THE LIVING WORLD, Belknap Press of Harvard University Press, Cambridge, Mass. (1997).

Palsson, B.O., "The challenges of *in silico* biology," *Nature Biotechnology*, **18**:1147–1150 (2000).

Price, N.D., Reed, J.L., and Palsson, B.O., "Genome-scale models of microbial cells: Evaluating the consequences of constraints," *Nature Reviews Microbiology*, **2**:886–897 (2004).

Wiechert, W., "Modeling and simulation: Tools for metabolic engineering," *Journal of Biotechnology*, **94**:37–63 (2002).

Yarmush, M.L., and Batna, S., "Metabolic engineering: Advances in the modeling and intervention in health and disease," *Annual Reviews in Biomedical Engineering*, **5**:349–381 (2003).

CHAPTER 13

Properties of Solution Spaces

As outlined in the previous chapter, we cannot uniquely compute the state of a biochemical reaction network. However, we can limit the range of possible functional states that a biochemical network can attain. With this range defined, there are two tasks that we can undertake. First, we can seek to characterize all the allowable states by using a set of basis vectors that can be used to span all the allowable solutions, and second, we can sample the solution spaces uniformly and determine the statistical properties of a large number of candidate solutions. These two approaches amount to an *unbiased* assessment of the properties of all the allowable states of a biological network.

13.1 Network Properties

Network functions arise from interactions of the components. Network properties and functional states are often referred to as *emergent properties* since they do not depend on the functions of any particular component, but "emerge" from the interactions of the components functioning together.

The pathway matrix
Since the extreme pathways (Chapter 9) are convex basis vectors and can represent all the functional states of a network, they can also be used to define, derive, and compute network properties. The *pathway matrix* $\mathbf{P}$ is formed using the extreme pathways ($\mathbf{p}_i$) as its columns:

$$\mathbf{P} = (\mathbf{p}_1, \ \mathbf{p}_2, \ \mathbf{p}_3, \dots) = (|||) \qquad (13.1)$$

Thus entries in row i of $\mathbf{P}$ indicate whether reaction i (v_i) is used in an extreme pathway. $\mathbf{P}$ can be written in a binary form ($\hat{\mathbf{P}}$), whose elements are defined by

$$\hat{p}_{ij} = 1, \ \text{if} \ p_{ij} \neq 0 \ \text{and} \ \hat{p}_{ij} = 0, \ \text{if} \ p_{ij} = 0 \qquad (13.2)$$

and they indicate whether a reaction is involved in the makeup of a pathway. This binary form of $\mathbf{P}$ is analogous to the binary form of $\mathbf{S}$ discussed in Chapter 7.

SVD of the pathway matrix

As for the stoichiometric matrix, SVD can be used to study some properties of $\hat{\mathbf{P}}$ [177]. SVD is described in detail in Chapter 8. It decomposes the extreme pathway matrix into three matrices,

$$\hat{\mathbf{P}} = \mathbf{U}\Sigma\mathbf{V}^{\mathrm{T}} \qquad (13.3)$$

The columns of $\mathbf{U}$ can be called *eigen-pathways* and the rows of $\mathbf{V}^{\mathrm{T}}$ can be called *eigen-participations* using an analogy to Figure 8.10. The eigen-participations, or the columns of $\mathbf{V}$, represent how the extreme pathways can be linearly combined to form the scaled modes of $\mathbf{U}$;

$$\hat{\mathbf{P}}\mathbf{V} = \Sigma\mathbf{U} \qquad (13.4)$$

Thus, the elements of $\mathbf{V}$ are the weightings on the pathways needed to reconstruct each of the modes of $\mathbf{U}$ as scaled by their respective singular values.

As with the stoichiometric matrix, SVD leads to conceptually useful quantities that give us information about properties of $\hat{\mathbf{P}}$. For instance, the singular value spectrum gives a measure of the effective dimensionality of $\hat{\mathbf{P}}$, and the modes give us the principle directions of variance in the set of extreme pathways [177, 178]. However, as we will see in the following sections, the adjacency matrices of $\hat{\mathbf{P}}$ are more useful.

Systems properties of interest

The pathway matrix can be used to define, compute, and study several useful properties of a network. We will discuss the following properties below:

- *Pathway length:* Classically, we think of a pathway as a linear sequence of events and the length of the pathway is the number of steps in this sequence. Extreme pathways can be much more complicated, and "length" becomes the number of reactions that participate in the pathway. This number is computed from an adjacency matrix of $\hat{\mathbf{P}}$.
- *Reaction participation:* The number of pathways in which a reaction participates is an important quantity. It indicates how many pathways

are affected if a reaction is removed from the network. The knock-out of a gene can lead to the removal of a reaction, or a reaction can be down-regulated. Reaction participation is computed from an adjacency matrix of $\hat{\mathbf{P}}$. More importantly, one can calculate *correlated subsets* of reactions that always appear together in the extreme pathways, effectively forming a network *module*. Thus, reaction participation is useful for the hierarchical analysis of networks.

- *Input–output relationships:* Type I extreme pathways contain primary exchange reactions, recall Figure 9.10A. These exchange reactions can be used to determine which output can be achieved from a given input. Conversely, one can determine all the input combinations that lead to a given output. Furthermore, one can determine which pathways have an identical set of inputs and outputs. The number of pathways with identical inputs and outputs give a measure of network, or *pathway redundancy*, by counting the number of different internal states that give the same external state. Finally, based on the input–output relationships, one can mathematically define *crosstalk* using the overlap between the inputs and outputs in a set of pathways.

- *Consequences of regulatory rules:* The reconstruction of regulatory networks can lead to a set of causal relationships that describe regulatory interactions. These logical statements can be such that an extreme pathway can never be expressed due to conflicts with the regulatory rules, or only expressed under certain environmental conditions. One can therefore get an assessment of how regulation reduces the allowable functional states.

- *Decomposition of a flux state into extreme pathways:* Convex basis vectors can be used to reconstruct any point in a convex space. As discussed in Chapter 9, this reconstruction is not unique. Thus, we may want to know all the ways in which a particular point in the space can be decomposed into constituent pathways. The α-spectrum is a method that helps to answer this question.

Example systems

Several different network properties are discussed in this chapter. To illustrate them, we use sample networks:

- A simple, easy to understand network is shown in Figure 13.1. It contains three extreme pathways. $ExPa_1$ and $ExPa_2$ are not simple linear reaction chains, but instead contain two outputs: E and the by-product. Extreme pathways can have any number of inputs or outputs. $ExPa_3$, like $ExPa_2$, maintains cofactor pools at steady state. Each of the extreme pathways results in the production of E.

Reaction network

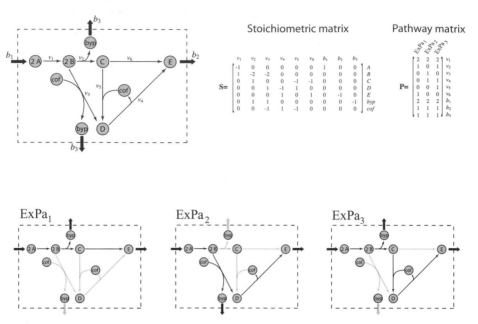

Figure 13.1: A simple network, its stoichiometric matrix, pathway matrix, and the flux map for the extreme pathways. From [162].

- The core metabolism in *Escherichia coli*. This network is described in Appendix B. Extreme pathways can be computed for growth on a variety of substrates. A summary of the extreme pathways found in this network is given in Table 13.1. The large number of extreme

Table 13.1: The number of extreme pathways found in the core *E. coli* model for four principal inputs (the columns; glucose pyruvate, succinate, and fumarate) and various outputs (the rows). Prepared by Ines Thiele.

Total number of ExPa	Glucose 613	Pyruvate 1749	Succinate 904	Fumarate 2598
Acetate out	—	464	232	591
2-oxoglutarate out	—	251	132	305
CO_2 out	613	1749	904	2598
Ethanol out	206	771	436	799
Formate out	—	892	620	1216
Proton in	—	649	820	2451
Water in	—	—	210	876
Water out	571	383	400	366
Lactate out	—	611	20	379
Oxygen in	571	1598	904	2036
Pyruvate out	—	—	40	58
Succinate out	—	508	—	1277

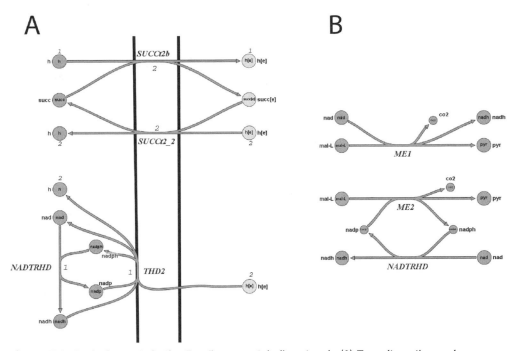

Figure 13.2: Equivalent sets in the *E. coli* core metabolic network. (A) Two alternative and equivalent ways to import a proton using two succinate transporters or two transhydrogenases. (B) Two equivalent ways to use the malic enzymes and transhydrogenase to produce pyruvate, CO_2, and NADH from malate and NAD. For each equivalent set there are two extreme pathways that are identical, except for the fluxes through the equivalent set. n equivalent sets representing two identical routes can lead to 2^n combinations of basically n different states. Network topology determines if they lead to n different states. Prepared by Ines Thiele.

pathways is due to the occurrence of *equivalent reaction sets* that leads to a combinatorial explosion; see Figure 13.2.

• The JAK-STAT signaling network in B cells has been reconstructed [160]. It accounts for 15 receptors and the corresponding ligands, and a total of 297 reactions (216 internal and 81 irreversible exchange reactions). A portion of the reaction map of this network is shown in Figure 13.3, and some of its finer details are illustrated using two representative sets of reactions in Figure 13.4. This set of reactions can be represented with a stoichiometric matrix and there are 147 extreme pathways found in this network [160].

13.2 Pathway Length

A pathway length matrix ($\mathbf{P}_{LM}$) can be calculated directly from the binary form of the extreme pathway matrix ($\hat{\mathbf{P}}$). $\mathbf{P}_{LM}$ is computed by premultiplying

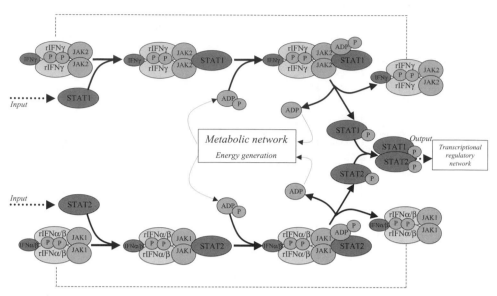

Figure 13.3: A portion of the reaction map for the JAK-STAT signaling network of the human B cell. Taken from [157].

$\hat{\mathbf{P}}$ by its own transpose,

$$\mathbf{P}_{\text{LM}} = \hat{\mathbf{P}}^{\text{T}}\hat{\mathbf{P}} \tag{13.5}$$

resulting in a symmetric matrix. $\mathbf{P}_{\text{LM}}$ is an adjacency matrix of $\hat{\mathbf{P}}$.

1. The diagonal values of $\mathbf{P}_{\text{LM}}$ correspond to the number reactions in an extreme pathway. In the example system (Figure 13.5), the first value along the diagonal is 6, meaning that six reactions participate in ExPa$_1$. A quick count of the fluxes shown in ExPa$_1$ (Figure 13.1) shows that there are indeed six reactions participating in the first extreme pathway.

2. The off-diagonal terms of $\mathbf{P}_{\text{LM}}$ are the number of reactions that a pair of extreme pathways have in common. For example, notice the circled diagonal term in Figure 13.5, which is a comparison of ExPa$_3$ (the column) and ExPa$_1$ (the row) and contains a value of 5. ExPa$_1$ and ExPa$_3$ have five reactions in common. Upon examining ExPa$_1$ and ExPa$_3$ in Figure 13.1, one can readily see that the five reactions shared are b_1, v_1, v_2, b_2, and b_3.

Thus, the diagonal and off-diagonal terms of $\mathbf{P}_{\text{LM}}$ are the number of reactions in an extreme pathway and the number of reactions common to the two pathways, respectively. The diagonal and off-diagonal elements in the matrix relate to the number of reactions in participating pathways and can be used to obtain a measure of pathway "length."

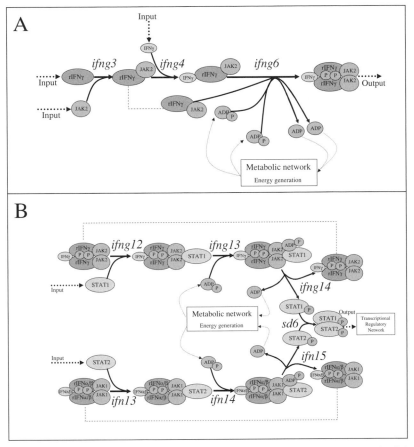

Figure 13.4: Representative sets of reactions in the JAK-STAT signaling network of the human B cell. The reactions in panel A correspond to the binding of the interferon γ ligand to its receptor and the subsequent formation of an activated receptor-ligand complex. The reactions in panel B correspond to the activation of the STAT1-STAT2 heterodimer by the interferon γ and interferon α/β bound receptors. Taken from [160].

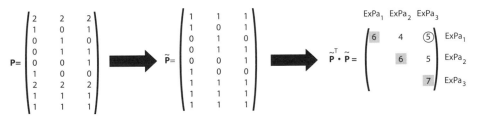

Figure 13.5: The pathway length matrix $\mathbf{P}_{LM}$ for the simple network in Figure 13.1. The lengths of $ExPa_1$, $ExPa_2$, and $ExPa_3$ are 6, 6, and 7, respectively, and are the highlighted diagonal elements of the final matrix. $ExPa_2$ and $ExPa_3$ have a shared length of 5, indicated by the circle. From [162].

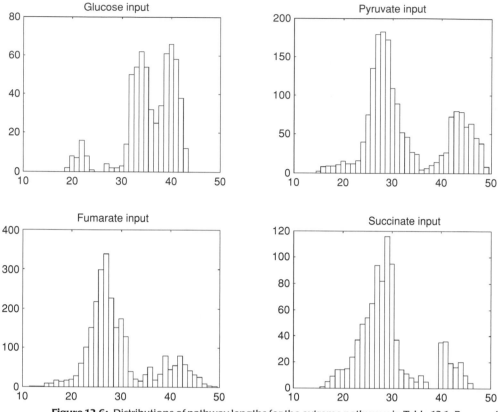

Figure 13.6: Distributions of pathway lengths for the extreme pathways in Table 13.1. Prepared by Ines Thiele.

E. coli core network

The pathway lengths for the extreme pathways in Table 13.1 can be computed using equation 13.5 and by studying the diagonal elements of $\mathbf{P}_{LM}$. The distributions of pathway lengths are shown in Figure 13.6 and are *bimodal*. For example, for pyruvate as an input, there are many pathways of lengths 25–35 and 42–48.

Genome-scale studies

Pathway lengths have been computed for genome-scale matrices [162]. For *Helicobacter pylori*, all the extreme pathways that lead to protein synthesis from a set of substrates have been computed (Figure 13.7). These pathways comprise about 100–110 reactions even though the protein yields vary significantly. This demonstrates that basically the same set of reactions can be used to generate quite different overall protein synthesis rates.

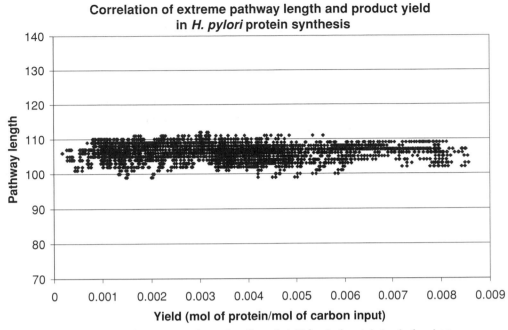

Figure 13.7: Correlation of extreme pathway length and yield (mol of protein/mol of carbon input) in *H. pylori* protein synthesis. There was essentially zero correlation between target product yield and extreme pathway length. From [162].

13.3 Reaction Participation and Correlated Reaction Subsets

The reaction participation matrix ($\mathbf{R}_{PM}$) is calculated by postmultiplying $\hat{\mathbf{P}}$ by its own transpose,

$$\mathbf{R}_{PM} = \hat{\mathbf{P}}\hat{\mathbf{P}}^{T} \qquad (13.6)$$

forming a symmetric matrix. $\mathbf{R}_{PM}$ is an adjacency matrix of $\hat{\mathbf{P}}$. The computed $\mathbf{R}_{PM}$ for the example system is shown in Figure 13.8.

1. The diagonal terms in $\mathbf{R}_{PM}$ give the number of pathways in which a particular reaction participates. For example, the first diagonal term in $\mathbf{R}_{PM}$ for the sample system, corresponding to reaction v_1, has a value of 3. Thus, reaction v_1 participates in all three extreme pathways. An examination of ExPa$_1$, ExPa$_2$, and ExPa$_3$ in Figure 13.1 shows that reaction v_1, which converts A to B, is in fact utilized in all three extreme pathways.

2. The off-diagonal terms give the number of extreme pathways that contains the pair of corresponding reactions. For example, the off-diagonal element boxed in Figure 13.8 has a value of 2. This element

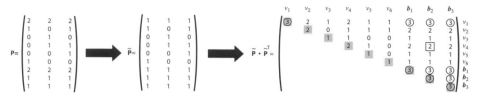

Figure 13.8: The reaction participation matrix $\mathbf{R}_{PM}$ for the simple network in Figure 13.1. The number of extreme pathways in which each reaction participates is indicated in the diagonal elements, as highlighted in the final matrix. These can be expressed as a percentage of the total number of extreme pathways. For example, reaction v_1 has a participation value of 3. Since there are three extreme pathways, this can be expressed as a 100% reaction participation. The off-diagonal terms can indicate correlated groups of reactions. Reactions v_1, b_1, b_2, and b_3 participate in three pathways. They also have a shared participation of 3, meaning they act as a correlated group, indicated by the circles. From [162].

refers to the number of pathways that contains both reaction b_2 (the column) and reaction v_4 (the row). Both of these reactions are utilized in ExPa$_2$ and ExPa$_3$, while only b_1 is utilized in ExPa$_1$ (Figure 13.1).

The values in $\mathbf{R}_{PM}$ can be converted to percentages by normalizing the entries in $\mathbf{R}_{PM}$ to the total number of extreme pathways, three in the example case. Thus, the first diagonal element would correspond to 100% reaction participation, since reaction v_1 was utilized in all three extreme pathways.

Reaction participation in the JAK-STAT signaling network

The reaction participation values have been calculated for the JAK-STAT signaling network (Figure 13.9). The reactions with participation values greater than 10% are shown in the inset table. Plotting the data on a log–log scale does not result in a linear relationship. Some of the observations from these computational results are:

- The exchange reactions of ATP and ADP have 100% participation. The conversion of ATP to ADP is essential for all the states of the signaling network since phosphate transfer is the mechanism by which the signal propagates. The signaling system is thus driven by the supply of ATP from energy metabolism.
- The exchange reactions for STAT1, STAT3, the STAT1-STAT3 heterodimer, and the reaction SD7, which describes the formation of the STAT1-STAT3 heterodimer, have the next highest participation values. The network is structured such that there are multiple routes to synthesize the STAT1-STAT3 heterodimer.
- There are 168 reactions that participate in only one extreme pathway. Reactions with low participation values correspond to reactions that

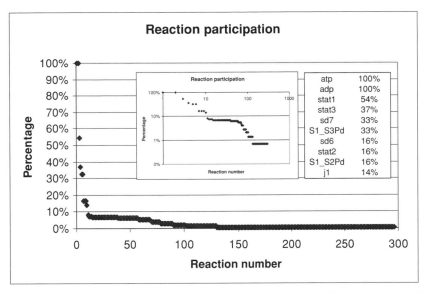

Figure 13.9: Reaction participation in the extreme pathways of the JAK-STAT signaling network. From [160].

have highly specific network functions. Manipulating reactions with low participation values could thus allow for control of very specific functions, potentially important for drug targeting.

Genome-scale studies

Reaction participation numbers have been computed for all the extreme pathways that lead to protein synthesis in *H. pylori*. The participation numbers for the individual reactions can be rank ordered (Figure 13.10) leading to the definition of three categories of reactions: (1) reactions that participate in all of the extreme pathways; (2) reactions that participate in varying amounts of extreme pathways; and (3) reactions that do not participate in any of the extreme pathways. The first group represents essential reactions for protein synthesis, and the last represents reactions irrelevant to protein synthesis. The second group represents a set of reactions that can be used for protein synthesis, but are not essential since there are pathways that lead to protein synthesis without them.

Correlated subsets

The off-diagonal elements of $\mathbf{R}_{PM}$ can be used to define correlated subsets of reactions [162, 166]. The circled elements in Figure 13.8 show reaction pairs that participate in exactly the same extreme pathways. In this particular case, each of these reaction pairs participates together in all of the extreme pathways. Thus, reactions v_1, b_1, b_2, and b_3 are always present.

**Reaction participation in *H. pylori*
amino acid synthesis**

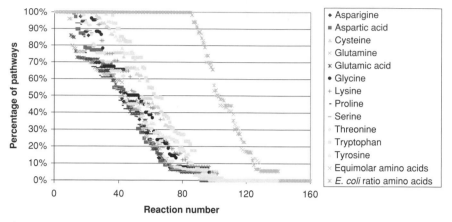

Figure 13.10: Reaction participation in the extreme pathways of the genome-scale *H. pylori* metabolic network. From [162].

They form a correlated reaction subset, meaning that if one of them is utilized, the others must also be utilized.

Correlated reaction subsets, *CoSets* [159], have been computed for genome-scale metabolic networks [162] and for the JAK-STAT signaling network. CoSets are informative with respect to the hierarchical organization of networks and are also useful for experimental design as discussed in Chapter 14.

CoSets in core *E. coli* metabolism

The CoSets for the extreme pathways in Table 13.1 have been computed (Table 13.2). They are shown graphically in Figure 13.11. Some observations from these results include:

- *CoSets for all substrates:* There are three CoSets of reactions that appear in the utilization of all the four substrates. CoSet 1 is reactions required for the secretion of ethanol; CoSet 2 is recycling of AMP; and CoSet 3 is the glyoxylate shunt in the TCA.
- *CoSets for three substrates:* There are seven CoSets that appear in the use of succinate, pyruvate, and fumarate, but not in the utilization of glucose. Two CoSets (number 4 and 11) are in the TCA cycle and CoSet 7 relates to the reverse use of glycolysis and the pentose pathway. These do not appear in the extreme pathways that use glucose as an input since glucose enters at the top of these pathways. CoSets 5, 6, 8, 9, and 10 relate to the secretion of acetate, α-ketoglutarate, formate, lactate, and water, respectively.

Table 13.2: The CoSets computed for extreme pathways in core *E. coli* metabolism (Table 13.1). Prepared by Ines Thiele.

CoSets	#	Members	Glucose	Found in Fumarate	Found in Succinate	Found in Pyruvate
All	1	ADHEr, ETOHt2r, EX_etoh	X	X	X	X
substrates	2	ADK1, PPS	X	X	X	X
	3	ICL, MALS	X	X	X	X
Three	4	ACONT, CS		X	X	X
substrates	5	ACKr, ACt2r, PTAr, EX_ac(e)		X	X	X
	6	AKGt2r, EX_akg(e)		X	X	X
	7	ENO, FBA, G6PDH2r, GAPD, GND, PGI, PGK, PGL, PGM, RPE, RPI, TALA, TKT1, TKT2, TPI		X	X	X
	8	FORt, PFL, EX_for(e)		X	X	X
	9	D-LACt2, LDH_D, EX_lac-D(e)		X	X	X
	10	H2Ot, EX_h2o(e)		X	X	X
	11	SUCOAS, TEST_AKGDH		X	X	X

Flux-coupling assessment through optimization

A more thorough assessment of the relationships between the use of fluxes in reconstructed networks has been developed [24], called the *flux coupling finder* (FCF). FCF is based on a linear programming approach to minimize and maximize the ratio between all pairwise combinations of fluxes in a reaction network. The computations classify reaction pairs to be:

1. *directionally coupled*, if a nonzero flux for v_i implies a nonzero flux for v_j, but not necessarily the reverse;
2. *partially coupled*, if a nonzero flux for v_i implies a nonzero, though variable, flux for v_j, and vice versa; or
3. *fully coupled*, if a nonzero flux for v_i implies not only a nonzero but also a fixed flux for v_j, and vice versa.

The FCF was utilized to analyze the genome-scale networks of *E. coli*, *Saccharomyces cerevisiae*, and *H. pylori* [24]. The percentage of reactions in the networks for each microorganism that were found in a coupled set was 60% for *H. pylori*, 30% for *E. coli*, and 20% for *S. cerevisiae*. This percentage is indicative of the flexibility of a network and the degrees of freedom available in a network.

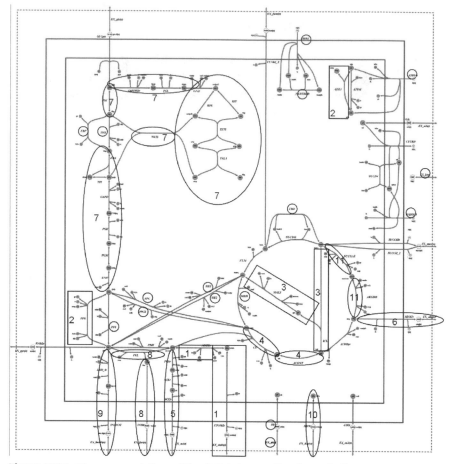

Figure 13.11: The representation of the CoSets in core *E. coli* metabolism on a reaction map. These sets are shown in Table 13.2. The CoSets in a box appear in the use of all four substrates in the table, while those in ovals appear in the use of only three of the substrates. The reactions in circles do not appear in any CoSet. The number in the CoSet is the same as the number in Table 13.2.

13.4 Input–Output Relationships

Extreme pathways have an input–output signature that is made up of the exchange reactions found in the pathway. These signatures can be used to generate an *input/output feasibility array* (IOFA).

The IOFA

The primary exchange reactions of an extreme pathway will contain input and output reactions (Figure 13.12). The columns of $\hat{\mathbf{P}}$ can be ordered based on the input and output status of a pathway. The extreme pathways with

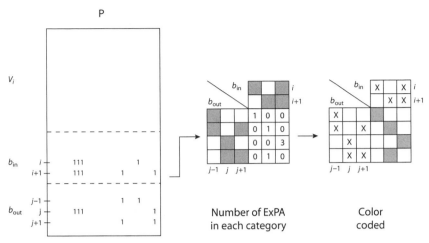

Figure 13.12: The input–output feasibility array.

identical inputs and outputs will be placed in an adjacent position, and
the number of such pathways enumerated.

The part of $\hat{\mathbf{P}}$ that contains the primary inputs and outputs can be segmented out of the matrix. This part of the matrix can be represented in
a different format. An array can be formed where the columns represent a unique set of inputs (the *input signature*) and the rows represent
a unique set of outputs (the *output signature*). If there is a pathway that
connects an input signature to an output signature the corresponding entry in the array can be assigned a value. This value can be a color so
that one can visualize all the possible matches between a set of inputs
and outputs. One can also put a numerical value in the array that corresponds to the number of pathways that connects a particular set of inputs
to a particular set of outputs. This array is called the IOFA. The IOFA
is a concise representation of a set of input–output properties of extreme
pathways.

The IOFA for the core *E. coli* network

The IOFA has been computed for the extreme pathways in Table 13.1; see
Figure 13.13. There are several interesting observations that follow from
these results. For instance, for fumarate and oxygen as the sole inputs,
there is only one output signature, formate and CO_2. The numbers in the
boxes give the number of extreme pathways that links the same sets of
inputs and outputs. This number is the number of different internal states
for a particular I/O relationship. For instance, for glucose as the sole input,
there are 42 ways to produce CO_2 and ethanol.

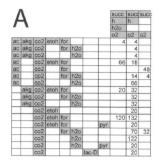
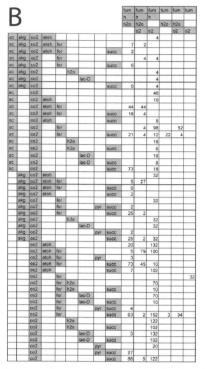

Figure 13.13: The IOFA for the extreme pathways in the core *E. coli* network. (A) Succinate as the primary input and (B) fumarate as the primary input. Prepared by Ines Thiele.

Computing the number of identical input/output states in a genome-scale metabolic network

Sets of extreme pathways for genome-scale metabolic networks for *H. pylori* and *Haemophilus influenzae* have been computed for a number of growth environments and for a number of required outputs (such as the production of individual amino acids). The average number of extreme pathways for all these different functional states that have identical input/output signatures has been computed [161]. The results show that for *H. pylori*, the number of pathways with identical inputs and outputs is 2, whereas the corresponding number for *H. influenzae* is 46. Thus, even though the metabolic networks appear similar, *H. influenzae* has much more flexibility in the choice of an internal state for a given overall network function.

13.5   Crosstalk

Definition

Extreme pathway analysis can be used to analyze the interconnection of multiple inputs and multiple outputs of signaling pathways, often called

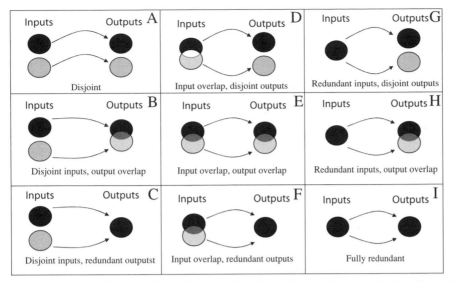

Figure 13.14: Classification scheme for crosstalk in the extreme pathways of a signaling network. From left to right, each pair of input sets is disjoint, overlapping, and identical. From top to bottom, each pair of output sets is disjoint, overlapping, and identical. From [161].

crosstalk. Since the extreme pathways are fundamental and irreducible functional states of a signaling network, crosstalk can be defined as the nonnegative linear combination of extreme pathways of a signaling network. The pair-wise combination of extreme pathways is thus the simplest form of crosstalk. The α-spectrum (see Section 13.7) could give a more complex breakdown of a measured functional state.

Classifying crosstalk

With this definition, crosstalk can be classified into nine different categories as shown in Figure 13.14. Each circle in Figure 13.14 represents a set of pathway inputs or a set of pathway outputs. From left to right, each pair of pathway input sets is classified as disjoint, overlapping, or identical. From top to bottom, each pair of pathway output sets is classified as disjoint, overlapping, or identical. For example, the representation in the middle of the figure corresponds to two extreme pathways with shared (but not identical) sets of inputs and shared (but not identical) sets of outputs. Thus, the two independent extreme pathways in this instance have overlapping but not identical functionality.

These nine categories provide a succinct description of fundamental properties of signaling networks:

- A pair-wise comparison of this type represents completely independent functions of a network (Figure 13.14A); completely separate inputs generate completely separate outputs.

Inputs \\ Outputs	●◐	●◐	●	Total
●◯	63.89	21.12	14.76	99.77
◐◯	0.00	0.00	0.00	0.00
●	0.05	0.15	0.04	0.24
Total	63.94	21.27	14.80	

Figure 13.15: Crosstalk analysis of the JAK-STAT signaling network following the classification theme in Figure 13.14. Taken from [160].

- Extreme pathways with disjoint inputs interact to generate overlapping outputs (Figure 13.14B). This type of interaction is conventionally thought of as crosstalk [201].
- Pair-wise pathway comparisons of the type in Figure 13.14C represent a nondiscriminate set of signals; two completely distinct inputs result in identical outputs. Recent experimental data show that cells can respond identically to different sets of stimuli.
- Interacting pathways of the type shown in Figure 13.14D would represent synergy or the use of co-signals; the network can use related inputs to generate distinct outputs.
- The case where a pair of pathways represents completely overlapping functions (Figure 13.14I) was discussed above and indicates that the same net result can be achieved in more than one way by the network.

The definition of these nine categories is possible using extreme pathway analysis and they allow for fundamental descriptions, such as those described above, of signaling networks and their functional states.

Crosstalk in the JAK-STAT signaling network

All nine categories of crosstalk have been computed for the reconstructed JAK-STAT network. With 147 extreme pathways, there are 10,731 ($= (147^2 - 147)/2$) pair-wise combinations. Approximately 99.8% of the pairs of extreme signaling pathways have disjoint outputs and nearly 0.2% have identical outputs. There are no pair-wise combinations of extreme pathways with overlapping outputs since all of the extreme pathways in the

JAK-STAT network have only one signaling output. Approximately 63.9%, 21.3%, and 14.8% of the pair-wise combinations have disjoint, overlapping, and identical input sets, respectively. The high percentage of pair-wise combinations with disjoint sets of inputs and disjoint sets of outputs indicates a fairly deterministic signaling network; there is very little classical crosstalk (i.e., identical signaling molecules used in different signaling pathways [201]) as one input signal typically corresponds to one output signal in this network.

13.6 Regulation and Elimination of Pathways

Regulation shrinks the solution space

All the allowable functional states of a reconstructed network are described by the corresponding set of extreme pathways. However, regulation may prevent the use of some of these functional states. Thus, regulation may be viewed as a way to shrink the steady-state flux solution space. This principle is illustrated in Figure 13.16. A flux cone is shown that has four extreme pathways. If regulation prevents the use of pathway 1, then the cone is reduced in size. Thus, regulatory networks can be viewed as a mechanism to generate self-imposed constraints that restrict the allowable functional states of a network. These constraints may be thought of as *restraints*.

Skeleton representation of the core metabolic pathways

A skeleton network of core metabolism has been formulated [38]. This network comprises 20 reactions, 7 of which are governed by regulatory logic. This network is a highly simplified representation of core metabolic processes, along with corresponding regulation, such as catabolite repression, aerobic/anaerobic regulation, amino acid biosynthesis regulation, and carbon storage regulation. A schematic of this skeleton network is shown in Figure 13.17, together with a table containing all of the relevant chemical reactions and regulatory rules which govern the transcriptional regulation. This network has 80 extreme pathways.

A total of 80 extreme pathways were calculated [36] for the simplified metabolic system shown in Figure 13.18. Whether a pathway is accessible to the cell depends on: (1) the regulatory network and (2) the environment in which the cell lives. Given the five inputs to the metabolic network and representing these inputs using Boolean logic (considering each as ON if present or OFF if absent), there are a total of $2^5 = 32$ possible environments, which may be recognized by the cell.

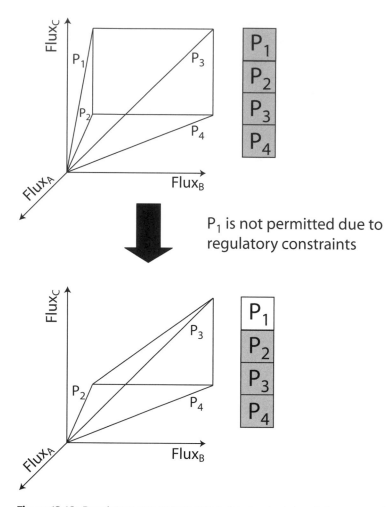

Figure 13.16: Regulatory constraints reduce the steady-state solution space of a metabolic network. A solution space bounded by invariant constraints on the network is shown. Extreme pathways may be calculated as the unique, systemically independent generating vectors for the space. In the space on top, all of the pathways are considered available to the system (denoted by the highlighted gray boxes at right). Under certain environments however, regulatory constraints may cause one or more of the extreme pathways to be temporarily unavailable to the system, P_1 in the case shown here (here ExPa$_i$ is denoted P_i). This results in a more restricted space with a reduced volume and/or dimension (bottom), corresponding to a metabolic network with fewer available behaviors. From [36].

Regulatory constraints alone eliminate 21 extreme pathways. The 32 environmental conditions allow anywhere from 2 to 26 extreme pathways to be expressed. Thus, the regulatory network for this simplified core metabolic model leads to a reduction of 67.5–97.5% in the number of available extreme pathways. To illustrate this process, a relatively simple dual-substrate environment is now described in more detail.

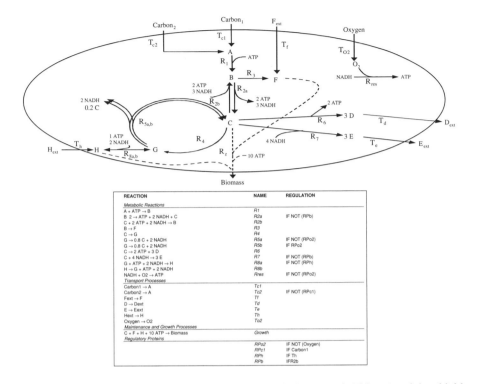

Figure 13.17: A schematic of the simplified core metabolic network. This network is a highly simplified representation of core metabolic processes, including a glycolytic pathway with primary substrates carbon1 (C1) and carbon2 (C2), as well as a pentose phosphate pathway and a TCA cycle, through which "amino acid" H enters the system. Fermentation pathways as well as amino acid biosynthesis are also represented. The regulation accounted for includes simplified versions of catabolite repression (e.g., preferential uptake of C1 over C2), aerobic/anaerobic regulation, amino acid (H) biosynthesis regulation and carbon storage regulation, and is also listed. The *Growth* reaction is indicated by a dashed line. From [36].

Growth on two carbon sources (C1 and C2) and oxygen (O_2)

If the core metabolic network operates in this nutritional environment and is able to secrete D and E while producing biomass, the number of allowable extreme pathways is substantially reduced (Figure 13.18). Initially, all 80 pathways are considered and are represented schematically in Figure 13.18A. Twenty-one of the extreme pathways are always restricted by regulation, as discussed earlier; the boxes representing these pathways are darkened in gray. By considering only the pathways with appropriate inputs and outputs based on the cell environment, 49 more pathways are eliminated (shaded in light gray). Of the 10 remaining pathways, six are inconsistent with the given regulation (C1 catabolite repression of the C2 transport protein Tc2 or regulation due to the aerobic environment) (shown in black) and the flux maps for the remaining four extreme pathways are shown in Figure 13.18B.

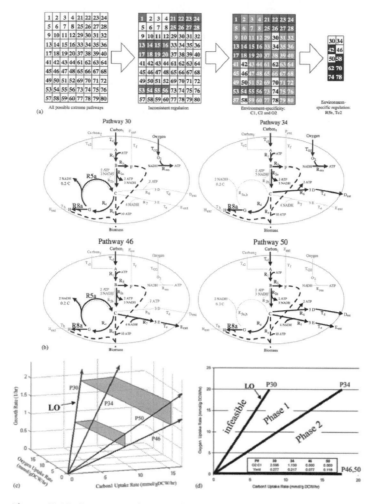

Figure 13.18: Extreme pathway reduction by constraints, using growth of the sample system in a C1 and C2 aerobic medium as an example. (A) The 80 total extreme pathways calculated for the system are represented by a grid. The number of pathways is reduced by 21 when pathways are removed that are always inconsistent with the regulatory rules (dark gray), then by 49 due to the specific environmental constraints (light gray), and then by 6 as the regulation corresponding to the specific environment is considered (black). The four remaining pathways, which are consistent with all the regulatory and environmental constraints, are shown schematically in (B), where the thick dark arrows represent active fluxes. (C) The solution space of the system, projected on a three-dimensional space, with the pathways and the line of optimality (LO) (the pathway with the greatest growth yield) noted. (D) A two-dimensional projection of the space, superimposed on a two-dimensional PhPP for the C1 and oxygen uptake. The region to the left of the LO lies outside of the space and is therefore infeasible. From [36].

The reduced solution space can be projected into three dimensions (represented by C1 uptake rate (x-axis), oxygen uptake rate (y-axis), and growth rate (z-axis)) showing the four feasible extreme pathways (Figure 13.18C). All the volume defined by these edges represents attainable functional states. The two-dimensional phenotypic phase plane (PhPP), see Chapter 16, for growth on C1 and O2 is shown in Figure 13.18D. This PhPP has two feasible phases between the lines shown, which represent projections of the four operational extreme pathways; pathways 46 and 50 are both fermentative and therefore overlap in the PhPP (oxygen uptake rate = 0). pathway 30 is the line of optimality as none of the carbon is lost in secretion of by-products; pathway 34 includes secretion of D and therefore gives a lower biomass yield than pathway 30.

Is the regulation of members of CoSets coordinated?

The CoSets that were discussed earlier in this chapter comprise reactions that appear together in all functional states of a network. Their regulation might therefore be coordinated. This possibility has been analyzed for the genome-scale *E. coli* model [184]. Two CoSets involving different carbon source usage during the optimal growth of *E. coli* are shown in Figure 13.19. They are computed from the metabolic network alone.

- Panel (A) in Figure 13.19 shows a CoSet containing four reactions involved in the utilization of the sugar rhamnose. The reactions in this CoSet are catalyzed by the products of four genes, which are organized into two operons. These operons form a regulon that is regulated by rhaS.
- Panel (B) in Figure 13.19 shows a larger CoSet that is involved in the degradation of the amino acid arginine. The 11 associated genes make up four operons. Currently, these four operons are not known to be regulated by a common regulator; however, the expression of the associated genes is found to be highly correlated ($P < 0.003$) across several growth conditions [184].

13.7 The α-Spectrum

The number of extreme pathways can exceed the dimension of the null space. Thus, there may not be a single set of weightings (α) on the extreme pathways that produce a given flux distribution, but rather a range of values. The numerical values of the weight on each extreme pathway in the reconstruction of a particular flux distribution can be determined using linear optimization to produce what is called the α-spectrum [244]. The α-spectrum defines which extreme pathways can and cannot be included in the reconstruction of a given steady-state flux distribution, and at what level [245].

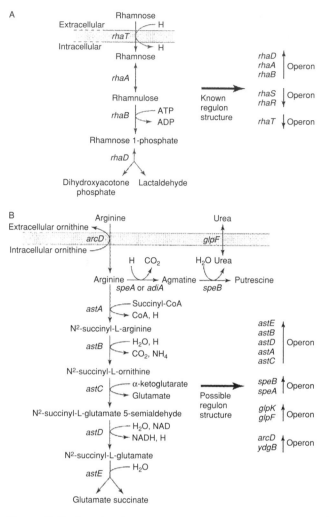

Figure 13.19: Correlated reaction sets and regulation. Taken from [159].

Defining the α-spectrum

The rank of $\mathbf{P}$ determines the number of independent equations, and is usually smaller than the number of extreme pathways, resulting in extra degrees of freedom. This results in an "α space" (Figure 13.20). In order to elucidate the range of possible α values that could contribute to the steady-state solution, the α-spectrum is computed based on the equation:

$$\mathbf{v} = \mathbf{P}\alpha \tag{13.7}$$

Where $\mathbf{P}$ is the pathway matrix, α is a vector of weightings on the extreme pathways, and $\mathbf{v}$ is a steady-state flux distribution. For each individual extreme pathway defined for the network, the α weighting for that pathway is both maximized and minimized using linear programming (see

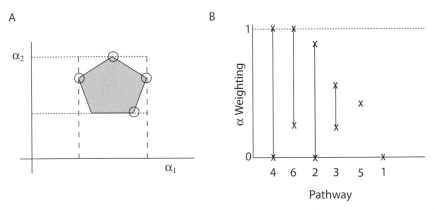

Figure 13.20: The α-spectrum. (A) A schematic graph of a two-dimensional α-space in which each α can vary from a minimum to a maximum forming an allowable range of α values (shown as shaded pentagon). The open circles represent the minimum and maximum possible values for each α. (B) A sample α-spectrum, rank-ordered according to the maximum α value with a corresponding pathway number on the x-axis. This sample α-spectrum shows all possible ranges of α: pathway 4: 0 to maximum usage (1 or 100%); pathway 6: a nonzero minimum to maximum usage; pathway 2: 0 to a submaximum usage; pathway 3: a nonzero minimum to a submaximum usage; pathway 5: a single nonzero value for α; and pathway 1: no usage of α (a single value of 0). Note that pathways 6, 3, 5 must be used in the reconstruction while pathway 1 cannot be used. Redrawn from [244].

Chapter 15) while leaving all other extreme pathway α weightings free. This computation will generate the range of weightings for every extreme pathway in the reconstruction of a particular flux distribution.

Computing the α-spectrum

The resulting mathematical formulation is summarized as follows:

Max α_i, subject to $\mathbf{v} = \mathbf{P}\alpha$, $i = 1, \ldots, n_p$, $0 \le \alpha_i \le 1$

Min α_i, subject to $\mathbf{v} = \mathbf{P}\alpha$, $i = 1, \ldots, n_p$, $0 \le \alpha_i \le 1$

where $\mathbf{v}$ is a particular flux distribution, the α_i-weightings correspond to each extreme pathway, $\mathbf{p}_i$, and n_p is the number of extreme pathways. Thus, $2n_p$ linear programming problems are solved to obtain $2n_p$ sets of α-weightings. Therefore, the maximum and minimum weighting of each pathway over all possible solutions to equation 13.7 can be obtained. This computation results in an allowable range for the α value for each extreme pathway.

The results are then plotted in a bar-type graph where the extreme pathway numbers are on the x-axis and their allowable ranges of α-weightings on the y-axis. The α_i value is normalized to the most constraining $v_{i,\max}$, and thus the α_i-weightings correspond to a percentage or fractional usage of each extreme pathway. Hence, the y-axis ranges from 0 (no usage) to 1 (100% usage). Some extreme pathways are not used while others can have a range of α_i-weightings (Figure 13.20). In some cases, the α_i-weighting is restricted to a single, nonzero value, meaning that particular extreme pathway is always utilized in that amount for the given flux distribution.

The conservative nature of the α-spectrum

As can be seen from Figure 13.20A, the defined α-spectrum is a rectangle that "boxes" in the range of allowable α values (as represented by the pentagon in Figure 13.20A). The rectangular shape is due to the fact that only one α-weighting is considered at a time; i.e., independent of the others. For a low-dimensional system, the true range of allowable weights fills a large fraction of the rectangle, but as the network size grows, this fraction becomes smaller and smaller. Thus, further refinement of the α-spectrum would be useful for genome-scale studies. One such refinement is to compute the minimum number of extreme pathways that was needed to describe a given flux distribution in cases where multiple extreme pathway combinations exist. Mixed integer linear programming (MILP) [249] is used for this purpose [244].

13.8 Summary

➤ Extreme pathways are a set of convex basis vectors that can be used to describe the entire steady-state flux solution space and can thus be used to characterize its content.

➤ The extreme pathways can be used as columns in a matrix to form a pathway matrix. This matrix can be represented in a binary form ($\hat{\mathbf{P}}$), where an entry of "1" indicates that a reaction participates in a pathway.

➤ The adjacency matrices of $\hat{\mathbf{P}}$ give the number of reactions that make up extreme pathways (pathway length), and the number of pathways in which a particular reaction participates (pathway participation).

➤ Analysis of pathway participation leads to the identification of correlated reaction sets (CoSets), which are sets of reactions that always appear together in the functional states of a network. The reactions that make up a CoSet may be coordinately regulated.

➤ The input/output status of pathways can be used to determine network redundancy and its crosstalk characteristics.

➤ Regulatory rules can be used to reduce the number of allowable pathways in a three-step process: first, the identification of pathways that are always in conflict with the regulatory rules, second, the identification of pathways that are inoperable in a given environment, and third, the identification of pathways that are rendered inoperable through regulation that is active in a particular environment.

➤ The decomposition of a particular steady-state flux vector into extreme pathways can be studied using the α-spectrum.

13.9 Further Reading

Papin, J.A., Price, N.D., Wiback, S.J., Fell, D.A., and Palsson, B.O., "Metabolic pathways in the post-genome era," *Trends in Biochemical Sciences*, **28**:250–258 (2003).

Papin, J.A., and Palsson, B.O., "The JAK-STAT signaling network in the human B-cell: An extreme signaling pathway analysis," *Biophysical Journal*, **87**:37–46 (2004).

CHAPTER 14

Sampling Solution Spaces

Extreme pathways are useful for studying the capabilities of a network and for determining network properties. Computing extreme pathways for genome-scale networks has proven difficult. An alternative approach to characterizing the contents of solution spaces is *random uniform sampling*. This approach involves obtaining a statistically meaningful number of solutions throughout the entire solution space and then studying their properties. A number of useful results can be obtained in this fashion. The sampling of solution spaces described in this chapter is an unbiased way of characterizing solution spaces.

14.1 The Basics

A simple flux split

Random uniform sampling can be illustrated by looking at a simple flux split; see Figure 14.1. Panel A shows a simple network that consists of one flux split, three exchange reactions, and constraints placed on the reactions. A three-dimensional space is formed by the three fluxes. The flux balance for this system is

$$-v_1 - v_2 + v_3 = 0 \tag{14.1}$$

or

$$\langle (-1, -1, 1) \cdot (v_1, v_2, v_3) \rangle = 0 \tag{14.2}$$

This equation defines a plane with the normal vector $\mathbf{n} = (-1, -1, 1)$. This plane intersects the positive orthant. The maximum and minimum constraints on the fluxes define a box, and the intersecting plane of equation 14.2 forms a closed space that is a segment of a plane (Figure 14.1B). This segment can be shown in two dimensions using its two conical basis

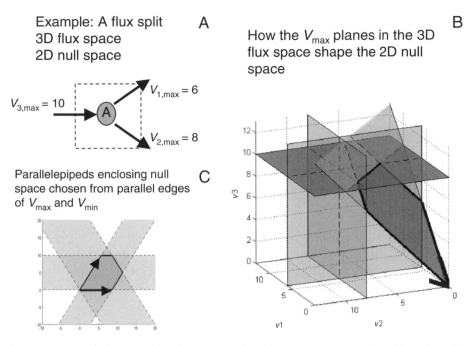

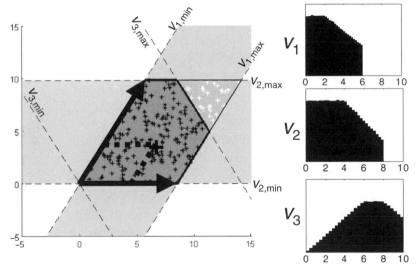

Figure 14.1: The simple flux split: the reaction network (A), overview of the process of forming the solution space (B), selecting one of the parallelepipeds (C), and the outcome and representation of randomized sampling. Taken from [179].

vectors (Figure 14.1C):

$$\mathbf{b}_1 = (1, 0, 1) \quad \text{and} \quad \mathbf{b}_2 = (0, 1, 1) \tag{14.3}$$

Note that both of these basis vectors are orthogonal to $\mathbf{n}$.

The plane segment can be enclosed with a parallelepiped, that is a parallelogram in two dimensions. The parallelepiped can then be sampled uniformly and a set of candidate solutions is formed by selecting only points in the segment (Figure 14.1D). The probability distributions for the flux through each individual reaction can then be computed. The algorithms used to perform the randomized sampling are discussed in the subsequent sections.

The overall procedure

The process of obtaining a uniform set of candidate solutions and studying their properties consists of three basic steps:

1. Defining the solution space based on imposed constraints;
2. Randomly sampling it based on uniform statistical criteria; and
3. Further segmenting the solution space based on additional post-sampling criteria as necessary.

It is particularly convenient to use only linear constraints in the first step. Linear equalities and inequalities lead to the formation of a polytope. Then, following the random sampling, candidate solutions can be eliminated based on nonlinear criteria in the third step, or based on additional experimental information. A large number of candidate solutions can then be characterized using statistical measures.

14.2 Sampling Low-Dimensional Spaces

To compute the size and contents of a solution space, Monte Carlo integration is generally implemented by defining a range of the variables that encompasses the solution space and then uniformly and randomly sampling points within this region. The solution space size is then calculated by determining the fraction of the uniformly distributed points that lie within the solution space and multiplying by the volume of the enclosing region, see Figures 14.1D and 14.7.

It is easy to sample regularly shaped geometric objects. A solution space can be enclosed by a geometric object in which uniformly distributed random points can be readily generated. Ideally, the shape of the chosen geometric object needs to fit as tightly as possible around the solution space. This would lead to a high fraction of points that are in the geometric object and in the enclosed solution space. This approach has been described in detail and the computational methods have been developed [179].

Parallelepipeds

A parallelepiped with the same dimension as the rank, r, of the null space of $\mathbf{S}$ is a geometric object that can enclose polytopes and can be readily sampled. A parallelepiped can be represented as a matrix, $\mathbf{B}$, where the columns of $\mathbf{B}$ represent a set of spanning edges of the parallelepiped,

$$\mathbf{B} = (\mathbf{b}_1, \ldots, \mathbf{b}_i, \ldots, \mathbf{b}_r) \tag{14.4}$$

Its volume is simple to compute [217]:

$$\text{Volume} = \sqrt{\text{Det}(\mathbf{B}^\mathsf{T}\mathbf{B})} \tag{14.5}$$

Sampling parallelepipeds

Uniform random samples of points can readily be generated within a parallelepiped. Uniform random weightings, α_i, are generated on all of the spanning edges, $\mathbf{b}_i$. A random point inside the space is generated by

$$\mathbf{v} = \sum_i \alpha_i \mathbf{b}_i, \quad \alpha_{i,\min} \leq \alpha_i \leq \alpha_{i,\max} \tag{14.6}$$

where $\mathbf{v}$ is a point within the space. This sampling procedure is readily applied to a simple flux split where there are only two basis vectors, as is illustrated in Figure 14.1D. Note that the maximum weight that can be placed on these vectors is $\alpha_{1,\max} = 6$ and $\alpha_{2,\max} = 8$, whereas the minimum values in both cases are zero.

Elimination of redundant constraints in determining $\alpha_{\max}$

Many reaction $v_{i,\max}$ levels cannot be reached in a steady state since the saturation of other reactions can be more constraining for v_i than its own $v_{i,\max}$ (recall Figure 9.6). Thus, many of the $v_{i,\max}$ constraints may be systemically redundant. Redundant $v_{i,\min}$ and $v_{i,\max}$ constraints are not needed to define the solution space. One can readily determine if a particular $v_{i,\max}$ is redundant by determining if

$$\max v_i < v_{i,\max} \quad \text{redundant} \tag{14.7}$$

or

$$\max v_i = v_{i,\max} \quad \text{not redundant} \tag{14.8}$$

along a spanning edge using the optimization methods described in the next chapter. Determining redundant $v_{i,\min}$ constraints is done similarly.

For the simple flux split in Figure 14.1, note that the maximum flux through reaction v_3 is redundant since the maximum fluxes on reactions v_1 and v_2 are more constraining on the spanning edges than that for reaction v_3. This fact is evident in Figure 14.1D where neither of the spanning edges reach the maximum value for v_3. The most constraining $v_{i,\max}$ of the reactions forming a spanning edge will determine the $\alpha_{i,\text{max}}$ value. Note that the $v_{3,\text{max}}$ constraint does become important after we sample the parallelepiped and it does lead to the exclusion of points from the random set of points obtained within the parallelepiped.

Choice of enclosing parallelepiped

Because each pair of $v_{i,\text{min}}$ and $v_{i,\text{max}}$ constraints form parallel hyperplanes, the shape of the null space leads naturally to the choice of a high-dimensional parallelepiped in which it will be enclosed. The set of possible parallelepipeds that can enclose the steady-state flux space is chosen by forming the faces of the parallelepiped along the directions defined by these $v_{i,\text{min}}$ and $v_{i,\text{max}}$ constraints. Since each parallelepiped is defined by r planes which are chosen from the set of m $v_{i,\text{min}}$ and $v_{i,\text{max}}$ planes, the number of such parallelepipeds that could be used to enclose the space is

$$\text{Number of possible parallelepipeds} = \binom{m}{r} = \frac{m!}{r!(m-r)!} \qquad (14.9)$$

where m is the number of v_{max} constraints and r is the dimension of the null space. For the simple split, $m = 3$ and $r = 1$; thus the number of possible parallelepipeds is 3. All three can be seen in Figure 14.1D.

The number of candidate parallelepipeds can be large, and it may be infeasible to compute the volumes for all of them. Therefore, an alternate approach is needed. One can choose the set of $v_{i,\text{min}}$ and $v_{i,\text{max}}$ constraints that are closest together based on Euclidean distance. Then, the second direction of the parallelepiped is chosen by determining the smallest parallelogram that can be formed by choosing the next set of $v_{i,\text{min}}$ and $v_{i,\text{max}}$ constraints. The third direction is then chosen as the set of constraints that forms the smallest parallelepiped using three sets of $v_{i,\text{min}}$ and $v_{i,\text{max}}$ constraints. This process continues until r sets of parallel planes are chosen and the so-formed parallelepiped fully encloses the solution space.

Uniform random sampling

A set of uniform random points can be generated within a solution space by randomly sampling within the enclosing parallelepiped. This can be performed by uniformly choosing a weight on each of the edges, b_i, of the parallelepiped. Each point in the space is uniquely defined by weightings on the edges spanning the parallelepiped (equation 14.6). The weighting, α_i, on each basis vector, b_i, can be uniformly selected by generating a random number, f, between 0 and 1 and computing each weight as

$$\alpha_i = \alpha_{i,\text{min}} + f(\alpha_{i,\text{max}} - \alpha_{i,\text{min}}) \qquad (14.10)$$

Points generated uniformly within the parallelepiped were then compared to the set of $v_{i,\max}$ and $v_{i,\min}$ constraints to verify whether or not the point falls in the solution space. If the point satisfies all constraints, it is a valid solution and is kept in the random set. If the point does not satisfy all the constraints, it is excluded.

Computing the "volume" of the solution space

We can determine the *hit fraction* ($\overline{p}$) as the ratio of sampled points that fall inside the solution space and n as the total number of sample points generated. The volume of the solution space can be calculated by multiplying the volume of the enclosing parallelepiped by the fraction of generated points that fall within the solution space:

$$\text{Estimated volume of solution space} \approx \overline{p} \times \text{volume}$$
$$\text{of enclosing parallelepiped}$$

Note that the notion of a volume here is not related to our usual definition of volume as the size of a three-dimensional solution space. For instance, the so-calculated volume of a 2D solution space is actually an area (i.e., see Figure 14.1D).

Error in computed volume estimate

The variance in the volume estimate is [173]

$$\sigma^2 = \frac{\overline{p}(1 - \overline{p})}{n} \leq \frac{1}{4n} \tag{14.11}$$

where n is the total number of sample points generated. The maximum variance is at $\overline{p} = 1/2$. The estimated relative error in the volume calculation, ϵ, can be calculated as the ratio of the standard deviation (σ) to the mean (μ) as

$$\epsilon = \frac{\sigma}{\mu} = \sqrt{\frac{\overline{p}(1 - \overline{p})/n}{\overline{p}^2}} = \sqrt{\frac{1/\overline{p} - 1}{n}} \tag{14.12}$$

Sampling high-dimensional spaces

The sampling methods described above have proven to work for polytopes of dimensions of up to 10–12. In spaces of higher dimension, more sophisticated sampling methods are required. The detailed description of these methods requires sophisticated statistics and is beyond the scope of this text. The interested reader can consult with some of the primary references [111, 125].

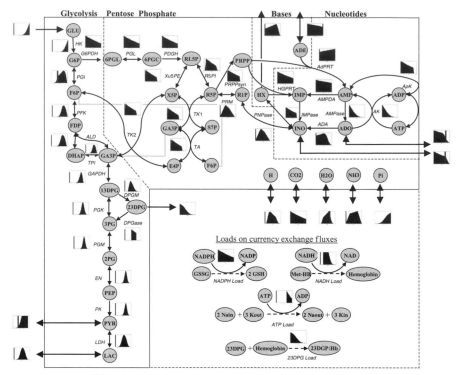

Figure 14.2: Monte Carlo sampling of the steady-state flux solution space for the red blood cell. From [179].

14.3 Applications to Biological Networks

Several studies have appeared that use random sampling of large solution spaces to study properties of reconstructed networks. This section describes some of these studies and the key results obtained to date. One can look forward to a productive use of this approach for genome-scale networks in the near future.

The red blood cell

The steady-state flux solution space of the human red blood cell has been studied through randomized sampling [179, 243]. The probability distributions for each flux in the network can be shown on the reaction map (Figure 14.2). This allows one to visualize all the allowable flux values for all the reactions in the network simultaneously. These studies have led to several notable results, three of which are briefly described here.

1. The histograms provide information about the "shape" of the solution space and how likely the fluxes are to fall into certain numerical values. For instance, some of the histograms are flat, implying

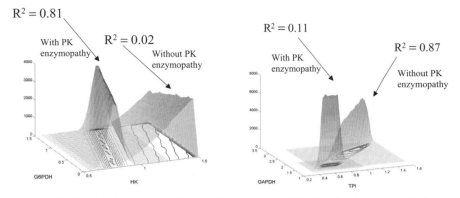

Figure 14.3: Simulated enzymopathies in the human red blood cell using segmentation of sampled solution spaces. From [179].

that every numerical value for a flux through a particular reaction is equally likely.

2. The cross-correlations between every pair of flux values can be computed. Such computations lead to identifying CoSets if the correlation coefficient is unity ($r^2 = 1.0$). The CoSets are also computable from the extreme pathways (Chapter 13). Less than perfect correlations can be found between flux variables, as described in more detail in the mitochondria example below. Such correlations can be used to guide experimental design. The measurement of poorly correlated fluxes is likely to be more informative than measuring highly correlated fluxes.

3. The pairs of fluxes that are correlated can be studied further. There may be regions in the solution space where they are strongly correlated and regions where they are poorly correlated. Such regions can be found by segmenting the solution space and computing the correlations in each segment; see Figure 14.4 for an illustration of this procedure.

   Solution spaces have been segmented based on inborn errors in metabolism, i.e., lowered $v_{i,\max}$ values due to genetic variation. Figure 14.3 shows an example of the analysis of enzymopathies in red cell metabolism. Pyruvate kinase (PK) and glucose-6-phosphate dehydrogenase (G6PDH) have many genetic variations in the human population, some of which are related to pathological conditions [103]. Figure 14.3 shows how the correlation between fluxes can be very different between the full solution space and the segment created by the imposition of a $v_{i,\max}$ constraint for PK and G6PDH. This suggests that the metabolic states could be significantly different in individuals with these enzymopathies as compared to a normal state.

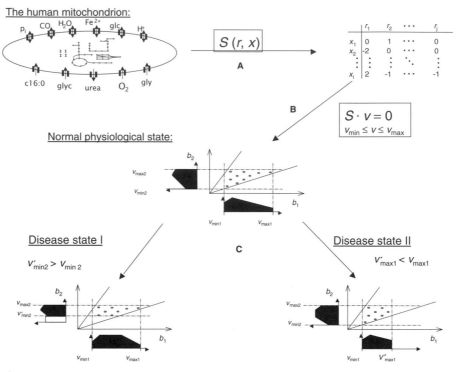

Figure 14.4: Sampling the steady-state flux solution space for the mitochondria in the human cardiac myocyte, and its segmentation based on restricting flux values. From [222].

The mitochondrion in human cardiomyocytes

The metabolic reaction network in the mitochondria in the human cardiac myocyte has been reconstructed based on proteomic data [238]. The steady-state flux solution space for this network has been sampled and the results used to study the effects of diabetes, ischemia, and dietary conditions. The overall procedure used in such a study is illustrated in Figure 14.4, following the three-step process outlined in the first section of this chapter. In the third step, altered $v_{i,\max}$ constraints are used to represent the disease and dietary conditions of interest.

A random sample of one-half million points was obtained for the mitochondrial metabolic network. All the pair-wise cross-correlation coefficients were computed and the CoSets identified. Such computations show the complicated and nonintuitive correlations between the various fluxes in the network; see Figure 14.5. Disease states were represented by segmenting the solution space based on measured changes in exchange rates, such as the uptake rates of nutrients and oxygen. The segmentation based on elevated fatty acid uptake (to simulate diabetic conditions) showed that the probability distribution for pyruvate dehydrogenase (PDH) becomes very narrow. Thus, network structure and allowable

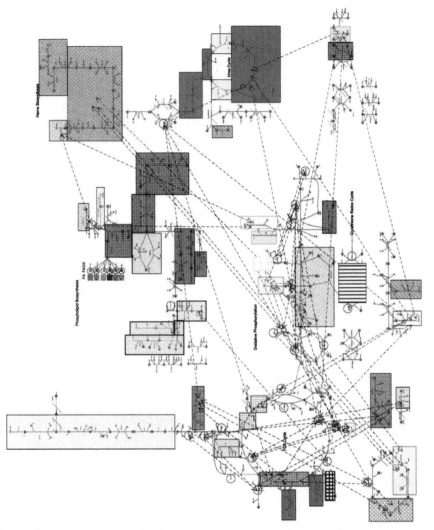

Figure 14.5: CoSets in metabolism in the human cardiomyocyte metabolism. The dashed lines indicate over 85% correlation between CoSets. From [222].

states under diseased conditions lead to a network structure-based restriction of the flux through PDH. This restriction is experimentally observed [44, 215, 219]. No unknown regulatory mechanism is therefore likely to be responsible for this observation.

Growth of *E. coli*

The steady-state flux solution space of a genome-scale reconstruction of *E. coli* metabolism has been randomly sampled [2]. The distribution of all the steady-state flux levels through all of the individual reactions in the sampled solutions shows an approximate power-law distribution; see Figure 14.6. This feature is a global property of the functional states of

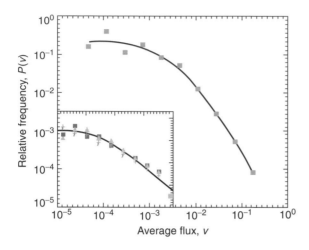

Figure 14.6: Distribution of the flux levels through reactions in a genome-scale metabolic network of *E. coli*. From [2].

the network. Recall from Chapter 7 that the distribution of the number of reactions that a compound participates in also follows an approximate power-law distribution.

This study also looked at the correlated use of reactions in the *E. coli* metabolic network. It was found that a large fraction of the candidate solutions had common network-scale patterns in the used sets of reactions. This large CoSet was termed the *high-flux backbone* of the flux map, since it was similar in all solutions. The individual solutions then have distinct deviations from this state.

14.4 Sampling the Concentration Space

The methods and studies described above are focused on the sampling of a solution space that contains the steady-state flux maps. The concentration spaces that are described in Chapter 10 are also polytopes that can be sampled using Monte Carlo methods. The additional constraints that can be stated on concentration spaces are nonlinear and lead to an additional step in the generation of candidate solutions for the concentration vector.

Defining the ranges of allowable values

Equation 10.6 defines a hyperplane that contains the concentration vector. Since the concentrations are positive variables, $x_i \geq 0$, the concentration solution space is the intersection of the hyperplane with the positive orthant. Since the definition involves linear equalities and inequalities, the minimum and maximum allowable concentration values ($x_{i,\min}$ and $x_{i,\max}$)

Figure 14.7: Monte Carlo sampling of a two-dimensional concentration space. The relative size of a constrained subspace, i.e., space defined by $x_{i,\min} \leq x_i \leq x_{i,\max}$, is computed by uniformly sampling the entire space and calculating the ratio of the points inside the constrained subspace over the total sampled points. Notice that one of the boundaries is nonlinear. Further segmentation based on an additional criterion for x_1 ($\leq x_c$) is illustrated. From [53].

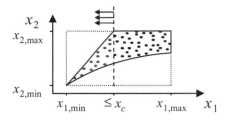

in this space can be found by linear optimization (see Chapter 15 for details):

$$\text{Max (or min) } x_i, \text{ subject to: } \mathbf{a} = \mathbf{Lx}, \quad \text{and } 0 \leq x_i, \quad \text{for all } i \qquad (14.13)$$

Imposing additional constraints on concentration spaces

When no constraints are imposed in addition to the equality and inequality constraints, the solution space size (volume) is considered to be 1, or 100%. Imposing additional constraints subsequently reduces the space size, thus giving a measure of *relative* size. The reduced space size is estimated simply by calculating what fraction (or percentage) of the random set falls within the constrained region (Figure 14.7). A calculated error in this estimate is obtained by computing the variance in estimating the size; see equation 14.11.

In addition to the mass conservation and nonnegativity constraints, there may be other factors further constraining the range of allowable concentration states. Thermodynamic constraints and measured concentrations further reduce the size of the concentration solution space:

- *Thermodynamic constraints* are imposed by including an inequality condition

$$\Gamma \leq K_{\mathrm{eq}} \qquad (14.14)$$

  where Γ represents the *mass action ratio*, which is the product to reactant concentration ratio. These constraints are typically bilinear.
- *Concentration constraints* based on measurements are imposed by constraining a compound concentration to within a range of the nominal value:

$$x_i = (1 \pm \epsilon)x_{i,\mathrm{measured}} \qquad (14.15)$$

  where ϵ represents the relative experimental error.

Normalized histograms and expected value calculation

The effect of the successive imposition of constraints can be visualized using a histogram of the randomized values for each concentration. To determine an expected value for each concentration in the solution space, the average value and the variance of the probability distribution can be evaluated and compared to experimental measurements.

Such histograms can be normalized in different ways. One way is to start with a certain number of points, and then eliminate points by imposing successive constraints. Alternatively, the histograms can be normalized so that the same number of points are used in each one. The number of points at each bin on the histogram represents the likelihood of observing that value when sampling each reduced subspace.

Specific cases

We will now consider the successive imposition of constraints over and above the linear equality and nonequality constraints used to define the hyperplane segment confined to the positive orthant. The two additional constraints considered are thermodynamically derived equilibrium constraints and concentration measurements. Furthermore, the allowable ranges of the values of the individual variables and their variance will be computed, as well as their expected values following the imposition of successive constraints.

Bimolecular association

The bilinear association reaction that converts C and P to form CP contains two conservation relationships and was described in equation 10.15. Note that the conservation relationships, i.e., the rows of $\mathbf{L}$, are independent of the equilibrium constant, K_{eq}, and result solely from the network topology. Assuming one known metabolite profile $x_{known} = (1, 1, 1)^T$, the conservation quantities can be calculated as $\mathbf{a} = (2, 2)^T$. The concentration solution space is thus a one-dimensional space residing at the intersection of the two planes formed by the basis of the left null space in the positive orthant; see Figure 14.8A.

For $K_{eq} = 0.22$, the nonlinear equilibrium constraint divides the solution space into two unequal regions. If the reaction proceeds in the forward direction, the mass action ratio, Γ, or product to reactant ratio, must be less than the equilibrium constant,

$$\Gamma = \frac{[CP]}{[C][P]} \leq K_{eq} \qquad (14.16)$$

and the constrained subspace lies at the lower side of the nonlinear constraint with 25% of the original size (Figure 14.8B).

If the reaction proceeds in a reverse direction, the nonlinear constraint is changed so

$$\Gamma = \frac{[CP]}{[C][P]} \geq K_{eq} \qquad (14.17)$$

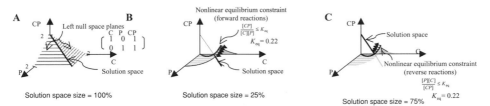

Figure 14.8: Concentration solution space reduction in bimolecular association schema. (A) Concentration space lies at the intersection of two planes formed by the basis vectors of the left null space. (B) The nonlinear equilibrium constraint segments the space into two regions. The lower segment forms the constrained region if the reaction proceeds in the forward direction (25% of the unconstrained space with $K_{eq} = 0.22$). (C) For the reaction occurring in the reverse direction, the upper segment defines where the constrained region resides (75% of the unconstrained space). From [53].

and the constrained subspace shrinks to 75% of the original (Figure 14.8C). The value of K_{eq} determines the percent of reduction in space size. For example, for an equilibrium constant of unity, the solution space will be divided into two equal regions of 50% each.

Thus, the imposition of equilibrium constraints may segment the solution space equally or unequally depending on the numerical value of the equilibrium constant. Note that the imposition of a nonlinear constraint in this example does not alter the linearity of the boundaries in the resulting subspaces, i.e., the reduced subspaces remain linear even though the additional constraint imposed is nonlinear.

Multicomponent cofactor coupled reactions

In Chapter 10 we studied multiple redox coupled reactions; see equation 10.4. A linear basis for the concentration space is given by

$$\mathbf{L} = \begin{pmatrix} 0 & 0 & 0 & 0 & 1 & 1 & 0 \\ 0 & 1 & 1 & 0 & 0 & 1 & 0 \\ 0 & 0 & 0 & 0 & 0 & 1 & 1 \\ 1 & 0 & 0 & 1 & 0 & -1 & 0 \end{pmatrix} \tag{14.18}$$

where the order of columns in $\mathbf{L}$ is RH_2, R, R, RH_2, H^+, NAD^+, and NADH. Assuming a concentration of 5 mM for each metabolite, the conservation quantities become $\mathbf{a} = (10, 15, 10, 5)^T$. With seven variables and four constraints, the concentration solution space is three dimensional, i.e., $7 - 4 = 3$.

The effect of equilibrium constants and concentration constraints can be observed in the three-dimensional space formed by RH_2, R, and NADH; see Figure 14.9. The nonlinear equilibrium constraints reduce the concentration space to a subspace with nonlinear boundaries that can be non-convex (Figure 14.9D). Note that the convexity of a constrained space is

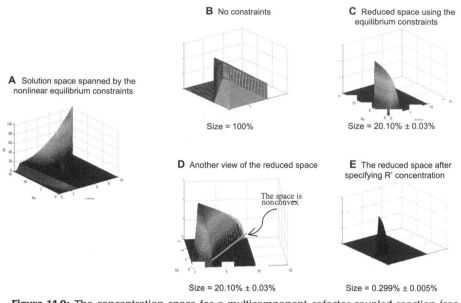

Figure 14.9: The concentration space for a multicomponent cofactor-coupled reaction (see equation 10.4) and its size reduction with the imposition of additional constraints. (A) The intersection of the two nonlinear spaces lies below the curved surface. (B) The solution space with no constraints is formed by the conservation relationships of $\mathbf{Lx} = \mathbf{a}$. (C) The solution space is reduced when the nonlinear constraints shown in (A) are applied. (D) Another view of (C) shows that the space is nonconvex, i.e., a line connecting two solution points lies outside the space. (E) The space is further reduced when a known metabolite concentration is specified as an additional constraint. From [53].

guaranteed only when the linear or nonlinear inequality constraints are also convex functions [15]. In this example, the equilibrium relationship of reaction 1 is nonconvex, and when imposed on the solution space results in the observed nonconvexity (Figure 14.9D). Thus, the imposition of nonlinear thermodynamic constraints can significantly change the characteristics of the solution space. The solution size can be further reduced if any of the internal metabolite concentrations are specified to a known value (Figure 14.9E).

14.5 Summary

➤ Solution spaces can be studied by randomly sampling points contained within them.

➤ Large sets of candidate solutions can be statistically analyzed.

➤ A three-step procedure comprising (1) confinement by linear constraints, (2) randomized sampling, and (3) confinement by nonlinear and other additional constraints can be implemented.

➤ Bounds and balances form linear constraints that comprise nonnegativity constraints on the variables and by mass or flux balances.

➤ Solution spaces can be enclosed by regularly shaped objects, such as parallelepipeds, making the sampling procedure easy and enabling the computation of the size (volume) of the solution space.

➤ Low-dimensional spaces (≤ 10–12) can be sampled by Monte Carlo based methods, whereas high-dimensional spaces require more sophisticated approaches.

➤ Metabolomic data, fluxomic data, and thermodynamic properties can be used to set additional constraints.

➤ The consequences of additional constraints can be assessed by determining the reduction in the size of the solution space.

14.6 Further Reading

Almaas, E., Kovacs, B., Vicsek, T., Oltvai, Z.N., and Barabasi, A.L., "Global organization of metabolic fluxes in the bacterium *Escherichia coli*," *Nature*, **427**:839–843 (2004).

Price, N.D., Schellenberger, J., and Palsson, B.O., "Uniform sampling of steady-state flux spaces: Means to design experiments and interpret enzymopathies," *Biophysical Journal*, **87**:2172–2186 (2004).

Thiele, I., Price, N.D., Vo, T.D., and Palsson, B.O., "Candidate metabolic network states in human mitochondria: Impact of diabetes, ischemia, and diet," *Journal of Biological Chemistry*, **280**:11683–11695 (2005).

Wiback, S.J., Famili, I., Greenberg, H.J., and Palsson, B.O., "Monte Carlo sampling can be used to determine the size and shape of the steady-state flux space," *Journal of Theoretical Biology*, **228**: 437–447 (2004).

Finding Functional States

For typical biological networks, the number of reactions (n) is greater than the number of compounds (m) resulting in a plurality of feasible steady-state flux distributions. Although infinite in number, the steady-state solutions lie in a restricted region, the null space (Chapter 9). The null space can be used to define the range of all allowable phenotypes of a given network [233], since it specifies all the steady-state flux distributions that it can achieve. However, only a particular set of phenotypes are expressed under particular conditions. Optimization can be used to find particular solutions of interest.

15.1 Finding "Best" Flux Distributions Through Optimization

The null space of the stoichiometric matrix is bounded since the fluxes have maximal values.[1] Linear optimization can be used to find solutions of interest within the bounded null space; see Figure 15.1. The bounded null space is defined by

$$\mathbf{S}_{\text{exch}} \begin{pmatrix} \mathbf{v} \\ \mathbf{b} \end{pmatrix} = \mathbf{0} \quad \text{where} \quad 0 \leq v_i \leq v_{i,\text{max}} \quad \text{and} \quad b_{i,\text{min}} \leq b_i \leq b_{i,\text{max}} \quad (15.1)$$

To pick out particular solutions within this space, one has to define the desired properties of such solutions. Mathematically, the definition of the solutions sought is stated in the form of an *objective function*. A general

---

[1] Note that a statement of maximal values for all fluxes may not be required. Recall Figure 9.6 and the associated discussion about redundant and dominant constraints.

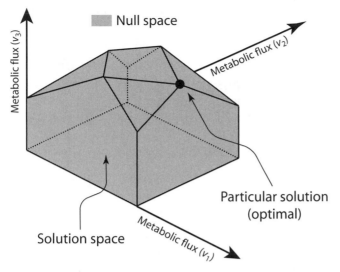

Figure 15.1: A schematic representation of the capped null space and a particular solution located within that space. Redrawn from [233].

linear objective function is defined as

$$Z = \mathbf{w} \cdot \begin{pmatrix} \mathbf{v} \\ \mathbf{b} \end{pmatrix} = \sum_i w_i v_i + \sum_j w_j b_j \qquad (15.2)$$

where the vector $\mathbf{w}$ is a vector of weights (w_i) on the internal and exchange fluxes, v_i and b_j respectively. The weights are used to define the properties of the particular solutions sought. Z is then optimized, i.e., minimized or maximized as appropriate. The solutions to these equations give the best use of the defined network to meet the stated objective function in a steady state. With a linear objective function (Z), this constrained optimization procedure is known as *linear programming* (LP). The constraints (equation 15.1) and the objective function (equation 15.2) can represent dual causation. The former is a statement of physicochemical constraints, while the latter can be used to represent biological features.

15.2 Objective Functions

The general representation of Z in equation 15.2 enables the formulation of a range of functionalities and network states of interest. Z can be used to represent exploration of the metabolic capabilities of a network, physiologically meaningful objectives (such as maximum cellular growth rate), or design objectives for a microbial production strain. A number of different objective functions have been used to analyze metabolic networks. These include:

- Minimize ATP production: This objective is stated to determine conditions of optimal metabolic energy efficiency and has been used to study the properties of the mitochondrion [181, 238].
- Minimize nutrient uptake: This objective has been used to study conditions under which the cell performs a particular metabolic function while consuming the minimum amount of available nutrients. This objective has been used to study yeast cultures grown in chemostats [54].
- Minimize the Manhattan (absolute) norm of the flux vector: This objective can be applied to satisfy the strategy of a cell to minimize the sum of the flux values, or to channel metabolites through the network using the lowest overall flux. Using the Euclidean norm of the flux vector would lead to *quadratic programming* (QP).
- Maximize metabolite production: This objective function has been used to determine the biochemical production capabilities of a particular cell, such as the maximal production rate of a chosen metabolite (i.e., lysine or phenylalanine). It also has been used to study the biochemical production capabilities of *Escherichia coli* [230]
- Maximize biomass formation: This objective has been widely used to determine the maximal growth rate of a cell in a given environment [49, 99]. We will discuss this objective function in detail in Section 15.5.
- Maximize biomass and metabolite production: By weighing these two conflicting objectives appropriately, one can explore the trade-off between cell growth and forced metabolite production in a production strain [25, 169, 170].
- More detailed objective functions that take thermodynamic and kinetic considerations into account have been described [92].

Some issues

The definition of the solution space has relatively few ambiguities associated with it, but the statement of the objective carries more uncertainties. There are a few important issues associated with the objective function of biomass formation:

- The biomass composition is variable and depends on the growth rate and the growth environment, and is different from one organism to another. These differences may change the predicted optimum behavior. This issue has been addressed [171, 172].

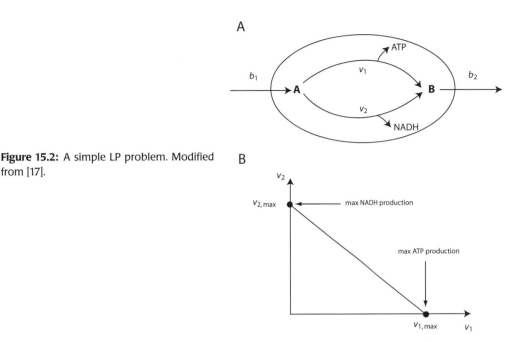

Figure 15.2: A simple LP problem. Modified from [17].

- Sensitivity calculations show that the optimum growth rates do not change significantly by varying the monomeric composition of the major macromolecules [233]. For instance, if the valine-to-alanine ratio is varied in the protein of a cell, the optimal growth rate does not significantly change. Conversely, if the protein relative to lipid composition in a cell changes, the optimum solution can be noticeably affected.
- The statement of a physiologically relevant objective function represents a guess about the "goals" of a cell. Although such assumptions can be rationalized, one never really knows the "true" objective. To help deal with this issue, one can invert this problem and look at an edge of the solution space, and then calculate all the objective functions that are maximized under those conditions [23].

15.3 Linear Programming: the Basics

Linear optimization is accomplished through LP. This optimization procedure is routinely used for the solution of a variety of different problems. The basics of LP are described in this section.

How LP works

A readily understandable example of an LP problem is shown in Figure 15.2. Panel A shows a reaction network where a compound A is picked

up by a cell and is metabolized to B via two different routes and then secreted. One route, v_1, produces high energy phosphate bonds in the form
of ATP. The other route, v_2, produces redox potential in the form of NADH.
The flux balance for this system is

$$v_1 + v_2 = b_1(=b_2) \qquad (15.3)$$

Since $v_1 \geq 0$ and $v_2 \geq 0$, these constraints define the solution space to be a
line segment that is the intersection of the positive orthant and the single
flux balance equation. Once b_1 is measured and has a known numerical
value, this intersection forms a closed line in the (v_1, v_2)-plane. Given a
stated objective, this line segment is searched for the best solution. The
optimal solutions for the maximization of ATP production or maximization
of NADH production are shown in Figure 15.2B and they lie at the ends
of the line segment that forms the solution space. This example shows a
one-dimensional solution space and the optimal solutions for two single-
valued objective functions.

Extreme points as optimal solutions
The fact that solutions lie at the edge of the allowable solution space is
particularly easy to see from the example in Figure 15.2. If one maximizes
ATP production, it is clear that v_2 should go to zero and v_1 to the maximum
value equal to the uptake rate. This optimal solution thus lies at the right
extreme point of the solution space. Conversely, if one maximizes the redox
production from this metabolite in the form of NADH, the optimal solution
is v_2 equal to the uptake rate b_1, and v_1 goes to zero. That optimal solution
is at the opposite end of the solution space.

Location of the optimal solutions
Next, we consider a slightly more complex example where a two-
dimensional solution space is formed by three inequalities (Figure 15.3).
We can also consider an objective function that is a combination of the two
variables. For a fixed value of Z, the objective function forms a straight line
in the two-dimensional plane. If the value of Z is changed, the line moves
and intersects the two-dimensional polytope at a different location. As we
increase the value of Z, the intersecting line moves closer and closer to
the periphery of the solution space. The maximum value for the objective
function is found when it intersects the solution space at a single point,
which is an extreme point in the space.

The types of solutions found
There are three types of feasible solutions encountered in solving LP problems. They are illustrated in Figure 15.4.

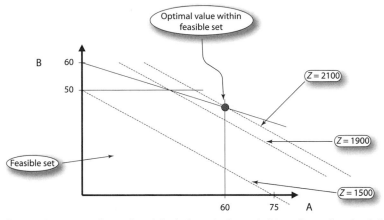

Figure 15.3: A two-dimensional depiction of a bounded two-dimensional solution space and an objective function that is being maximized. In this illustrative case, there are no equality constraints, but there are inequality constants: $0 \geq A \geq 60$ and $0 \geq B \geq 50$. Additionally, there is a simultaneous constraint: $A + 2B \leq 120$. The objective function is $Z = 20A + 30B$.

1. *Unique solutions:* For small networks, the optimal solution typically lies at an extreme point of the feasible set, as is the case in Figure 15.3.
2. *Degenerate solutions:* In some instances, the line formed by a constant value of an objective function is parallel to a constraint. In this case, the entire edge of the feasible set has the same value as the objective function, and all the points along the edge represent an optimal solution. This edge represents an infinite number of solutions, and mathematically they are called degenerate solutions. These solutions are *equivalent* optimal solutions since they correspond to the same value of the objective function. The occurrence of equivalent optimal solutions is frequent in genome-scale networks [184], and thus genome-scale networks are typically able to achieve the same overall functional network state in many different ways.

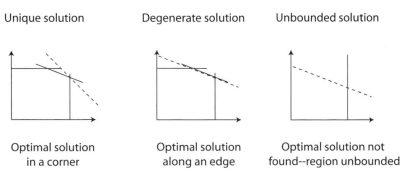

Figure 15.4: A graphical representation of the types of feasible solutions found by LP. The dashed lines are lines of constant Z.

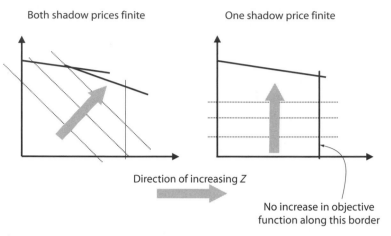

Both shadow prices finite

One shadow price finite

Direction of increasing Z

No increase in objective
function along this border

Figure 15.5: A graphical representation of a zero and nonzero shadow price at the edge of a boundary. The thick lines denote constraints, while the dashed lines represent lines of constant value of the objective function as in Figure 15.3.

3. *Unbounded solutions:* Sometimes the feasible set is unbounded and the objective function increases without limit in the open direction. In this case, no solution is found. Biologically, such situations are unrealistic, and if detected, typically result from an incomplete network formulation.

If the constraints are inconsistent, then the set of feasible solutions is empty and no solution can satisfy the stated constraints. In such cases, the constraints are incorrectly formulated.

Assessment of the sensitivity of the optimum solution

The sensitivity of the optimal solution is measured by *shadow prices* and *reduced costs*.

- *Shadow prices:* The shadow prices are the derivatives of the objective function at the boundary with respect to an exchange flux:

$$\pi_i = -\frac{\partial Z}{\partial b_i} \qquad (15.4)$$

The shadow prices define the incremental change in the objective function if a constraining exchange flux is incrementally changed. Shadow prices may change discontinuously as b_i is varied. The shadow prices can be used to determine whether an optimal functional state of a network is limited by the availability of a particular compound (Figure 15.5). The shadow prices thus essentially define the intrinsic value of the metabolites toward attaining a stated objective. This feature has proven useful for interpreting optimum solutions and for metabolic decision making [231]. We note that in some literature, the definition of a shadow price is the negative of what is stated in equation 15.4.

- *The reduced costs:* The reduced costs can be defined as the amount by which the objective function will change with the flux level through an internal flux that is not in the basis solution (i.e., fluxes that have a zero net flux):

$$\rho_i = -\frac{\partial Z}{\partial v_i} \qquad (15.5)$$

Several important questions arise that can be addressed using the reduced costs. The reduced costs can be used to analyze the presence of alternate equivalent flux distributions. If a reduced cost is zero, it means that the flux level through the corresponding reaction does not change the objective function. Thus, the reduced costs can be useful examining the effect of gene deletions on the overall function of a network.

15.4 Exploring Network Capabilities

The linear programming methods that have been outlined and illustrated for very simple examples are highly scalable and can be applied to networks of arbitrary complexity. In this section, we will perform a few illustrative computations using the core *E. coli* network detailed in Appendix B. These methods are now routinely applied at the genome scale [110, 175].

The core metabolic network in *E. coli*

The core metabolic network of *E. coli* (see Appendix B) contains 56 compounds and 64 reactions. This network can be used to compute various properties of *E. coli* metabolism. The solution to the linear optimization problem is a 64-dimensional flux vector **v**, a string of 64 numbers. These numbers are hard to interpret directly. The best way to interpret the solution is to put the numbers on the reaction map so that the steady-state flux distribution can be visualized in the form of a flux map. The reaction map for the core *E. coli* metabolism is shown in Appendix B.

Determining network properties

The core *E. coli* metabolic network can be interrogated for various capabilities, including the maximal yields of cofactors, precursors, and biomass

Table 15.1: Maximum stoichiometric yields of biosynthetic precursors from glucose under aerobic conditions. Note that the production of two OAA requires eight carbon molecules, two of which come from CO_2. The core network was allowed to fix CO_2 during these computations.

Metabolite	Maximum yield	π_{ATP}	π_{NADH}	π_{NADPH}
G6P	0.889	−0.042	0.069	−0.069
F6P	0.889	−0.042	0.000	−0.069
R5P	1.055	−0.049	0.000	−0.082
E4P	1.297	−0.061	−0.101	−0.101
G3P	1.684	−0.079	−0.101	−0.132
3PG	2.000	0.000	0.000	0.000
PEP	2.000	0.000	0.000	0.000
PYR	2.000	0.000	0.000	0.000
AcCoA	2.000	0.000	0.000	0.000
αKG	1.000	0.000	0.000	0.000
OAA	2.000	0.000	0.000	0.000

achievable from a particular substrate or in a particular growth environment.

- *Yield of key cofactors:* If the core *E. coli* metabolic network is provided one unit of glucose and unlimited oxygen, it will maximally produce 15.00 units of ATP. This number is different than the number (38) reported in standard biochemistry textbooks. The difference is due to the fact that the electron transport system (ETS) in *E. coli* does not have a P/O ratio of 3 as assumed to be the case for mitochondria in animal cells. The P/O ratio for *E. coli* is closer to 1.0. The flux distribution that leads to maximal ATP production is shown on the reaction map in Figure 15.6A. The production and use of ATP during maximal production is shown in Figure 15.6B. Note that PFK uses one ATP, whereas first phosphorylation step to G6P is performed by the PTS by transferring the high energy phosphate group from PEP. Figure 15.7 shows the maximal production rates of NADH from glucose.
- *Yield of the biosynthetic precursors:* Various yields can be calculated based on any carbon source, once the LP problem has been set up. The results from maximal yield computations of the biosynthetic precursors using one unit of glucose as an input are shown in Table 15.1. The shadow prices can be used to determine the governing constraint, such as additional needs for key cofactors.

A

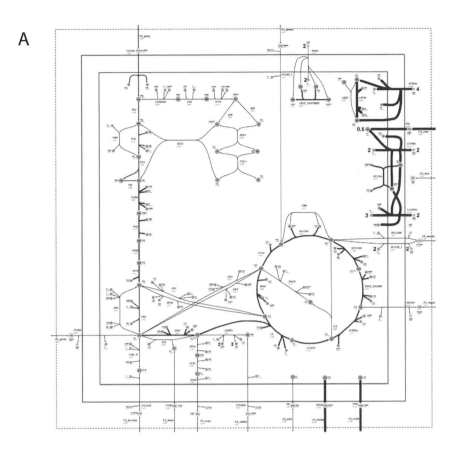

B

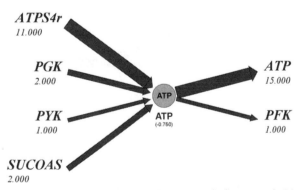

Figure 15.6: Maximal ATP production in the *E. coli* core metabolic network. Maximum ATP yield is 15 mmol/mmol glucose. (A) The flux map. (B) The rates of production and use of ATP under maximal yield conditions. Prepared by Adam Feist.

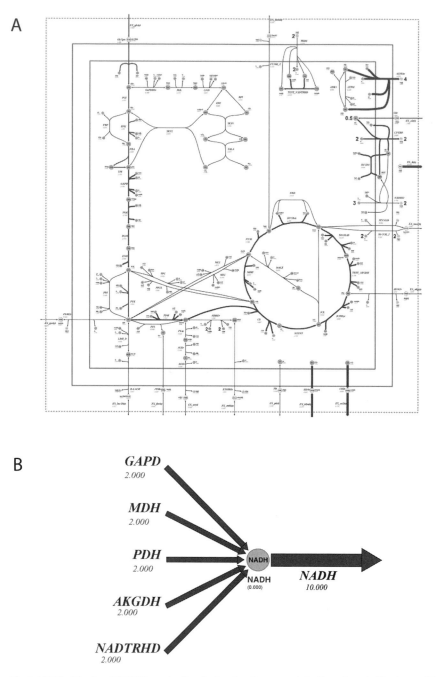

Figure 15.7: Maximal NADH production in the *E. coli* core metabolic network. Maximum NADH yield is 10 mmol/mmol glucose. (A) The flux map. (B) The rates of production and use of NADH under maximal yield conditions. Prepared by Adam Feist.

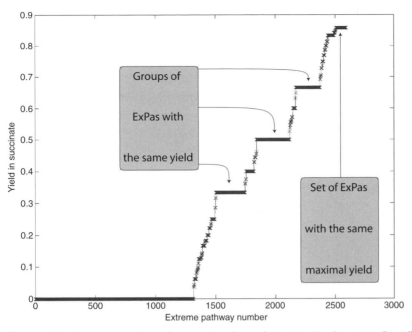

Figure 15.8: The generation of succinate from fumarate in the core *E. coli* model. The molar yield (mol/mol) of all 2,598 extreme pathways is shown. Prepared by Ines Thiele.

Occurrence of equivalent solutions

The relatively small core *E. coli* network has equivalent optimal solutions. The same optimal production of succinate from fumarate can be achieved in more than one way (Figure 15.8). The optimal value for succinate production is 0.86 mol/mol. This identical value can be achieved in 88 different ways; see Figure 15.8. In other words, the core network can support the same rate of succinate production through the use of different functional states. Two examples are shown in Figure 15.9, where two very different flux distributions result in the same overall yield.

Extreme pathways and optimal states

The extreme pathways are extreme conditions in the flux solution space and can thus represent optimal solutions. The extreme pathways that give a succinate yield of 0.86 from fumarate are alternate equivalent optimal solutions. In fact, any nonnegative linear combination of them (with $\Sigma\alpha_i = 1$, see Section 13.7) will also be an optimal solution. Figure 15.8 shows all the extreme pathways in the core model that lead to biomass generation, rank ordered by yield.

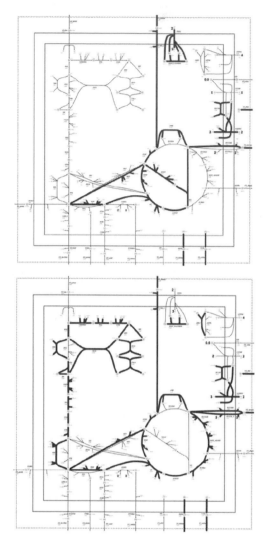

Figure 15.9: Two flux maps that give the same succinate generation from fumarate of 0.86, but have very different flux distributions. Prepared by Ines Thiele.

15.5 Producing Biomass

Growth can be defined in terms of the biosynthetic requirements to make a cell. These requirements are based on literature values of experimentally determined biomass composition. Thus, biomass generation is defined as a linked set of reaction fluxes draining intermediate metabolites in the appropriate ratios and represented as an objective function Z. This concept is illustrated in Figure 15.10A, which shows a schematic of a cell where a variety of substrates can enter the cell to produce all the compounds that constitute cellular components.

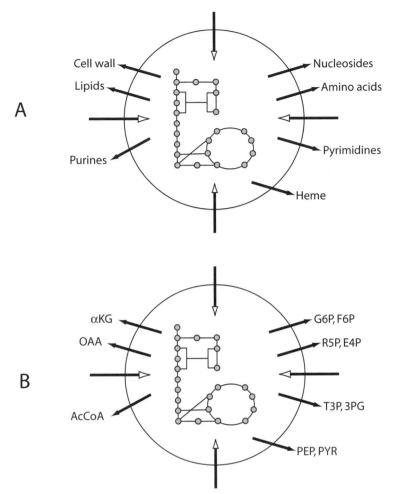

Figure 15.10: Biomass synthesis as an optimization problem. (A): Different inputs can be used to lead to a series of linked outputs that lead to the generation of all the cellular constituents. Some objective functions are used for genome-scale models. (B): The core metabolic network of *E. coli* and the production of biosynthetic precursors. These precursors need to be produced in the ratios given in equation 15.7.

Biomass formation in *E. coli*

The requirements for making 1 g of *E. coli* biomass from key cofactors and biosynthetic precursors have been documented [143]. This means that for *E. coli* to grow, all these components must be provided in the appropriate relative amounts. Key biosynthetic precursors are used to make all the constituents of *E. coli biomass* [143]. Their relative requirements to make 1 g of *E. coli* biomass are

$$
\begin{aligned}
Z_{\text{precursors}} = {} & +0.205 v_{\text{G6P}} + 0.071 v_{\text{F6P}} + 0.898 v_{\text{R5P}} \\
& + 0.361 v_{\text{E4P}} + 0.129 v_{\text{T3P}} + 1.496 v_{\text{3PG}} \\
& + 0.519 v_{\text{PEP}} + 2.833 v_{\text{PYR}} + 3.748 v_{\text{AcCoA}} \\
& + 1.787 v_{\text{OAA}} + 1.079 v_{\alpha\text{KG}}
\end{aligned}
\tag{15.6}
$$

The biomass composition of a cell thus serves to define the weight vector **w** in this objective function. In addition to the material that is needed to form biomass, cofactors are needed to drive the process. The cofactor requirement to synthesize the monomers from the precursors (amino acids, fatty acids, nucleic acids) and to polymerize them into macromolecules is

$$Z_{\text{cofactors}} = 42.703 v_{\text{ATP}} - 3.547 v_{\text{NADH}} + 18.22 v_{\text{NADPH}} \qquad (15.7)$$

Note that the biosynthetic reactions generate net redox potential in form of NADH. Thus the mass and cofactor requirements to generate *E. coli* biomass are:

$$Z_{\text{biomass}} = Z_{\text{precursors}} + Z_{\text{cofactors}} \qquad (15.8)$$

The full growth function for a genome-scale network of *E. coli* is more complicated than the one given above [185].

Maintenance energy requirements

To simulate growth situations, the biomass maintenance requirements have to be accounted for. Energy is used for both growth associated and nongrowth associated maintenance functions. These requirements for *E. coli* are estimated to be 7.6 mmol/ATP/gDW/h and 13.0 mmol ATP/gDW respectively [229]. The latter represents use of ATP that is proportional to the biomass being produced, while the former is constant drain that needs to be satisfied even in the absence of growth. Thus, the maintenance requirements are

$$Z_{\text{maintenance}} = Z_{\text{growth associated}} + Z_{\text{nongrowth associated}} \qquad (15.9)$$

and the combined biomass synthesis and maintenance requirements form a demand function

$$Z_{\text{growth}} = Z_{\text{biomass}} + Z_{\text{maintenance}} \qquad (15.10)$$

that requires both functions to be fulfilled at the same time.

The flux distribution through the core *E. coli* metabolic network for the optimal growth computed based on equation 15.10 is shown in Figure 15.11. The growth rate achieved is 0.59/h. If the maintenance requirements are removed from the objective function, and we recompute the growth rate based on equation 15.8 we get a growth rate of 0.64/h. The flux distribution changes a little bit (no flux changes by more than 2 mmol/gDW/h). Using $Z = Z_{\text{growth}}$, less flux goes through the pentose pathway and less NADPH for biomass synthesis is available. The flux down glycolysis through the TCA cycle increases (relative to $Z = Z_{\text{biomass}}$) that

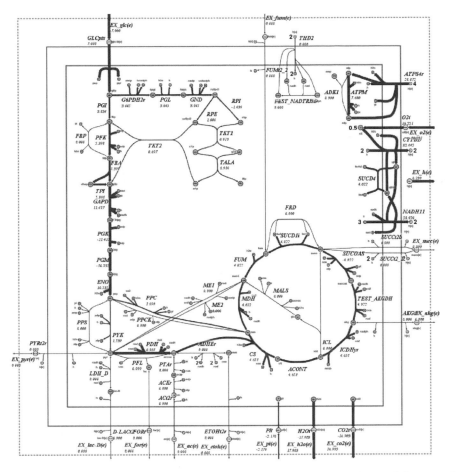

Figure 15.11: The flux distributions for the core metabolic network in *E. coli* for growth (biomass production plus maintenance, equation 15.10) from an input of 7 mmol glucose per gDW per hour. The thickness of the arrows is proportional to the flux through the reaction. The computed growth rate is 0.59/h. Prepared by Jennie Reed.

leads to more production of NADH whose electrons go down the ETS to produce more ATP to meet the maintenance requirements.

The growth rate can be computed as a function of the glucose consumed, Figure 15.12. The fulfillment of the stoichiometric biomass synthesis requirement is a straight line through the origin. Once the two maintenance requirements are added, the line intersects the *x*-axis and becomes slightly nonlinear. The glucose uptake rate at intercept with the *x*-axis represents the glucose that must be consumed just to satisfy the nongrowth associated maintenance requirement. The growth rate increases more slowly with the glucose uptake rate when the growth associated maintenance requirement is imposed since not all the glucose consumed can be used to meet biosynthetic requirements.

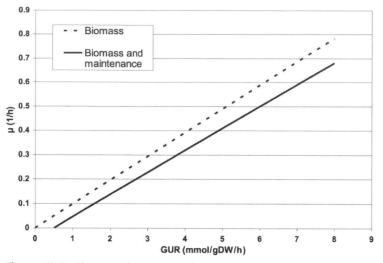

Figure 15.12: The growth rate as a function of glucose uptake rate. The two cases correspond to just biomass synthesis (equation 15.8) and to biomass synthesis with simultaneous fulfillment of the maintenance requirements (equation 15.10). Prepared by Adam Feist.

Gene deletions

One common application resulting from the ability of LP to determine network capabilities has been to assess the consequences of gene knockouts. If a gene is removed from the genome of an organism, one or more reactions can be rendered inoperative. The GRP associations (see Chapter 3) determine how a gene deletion affects the reactions in the network. The capabilities of the network without these reactions can be assessed and compared to the full network using the same procedures as discussed above. Gene knockout strains and their growth properties have been studied at the genome scale. An increasingly large set of knockouts and growth conditions have been performed in recent years [71, 115]. For *Saccharomyces cerevisiae* and *E. coli* these results show that genome-scale models compute the consequences of metabolic gene knockouts correctly in about 70–90% of the cases considered.

A simple example of a gene deletion analysis is the determination of the effects of the removal of each gene in the core *E. coli* model on the network's ability to generate the biosynthetic precursors. Such an analysis is performed by setting the objective function to maximize the production of one of the biosynthetic precursors. Then one performs repeated optimization computations in each of which a flux is constrained to zero for a reaction that is rendered inactive due to gene deletion. The procedure is then repeated for all the biosynthetic precursors of interest.

Reaction Deleted	Biomass Yield (g DW/mmol glucose)	Precursor Yields										
		3pg	accoa	akg	e4p	f6p	g3p	g6p	oaa	pep	pyr	r5p
ACKr												
ACONT												
ACt2r												
ADHEr												
ADK1												
AKGt2r												
ATPS4r												
CO2t												
CS												
CYTBD												
D-LACt2												
ENO												
ETOHt2r												
FBA												
FBP												
FORt												
FRD												
FUM												
FUMt2_2												
G6PDH2r												
GAPD												
GLCpts												
GND												
H2Ot												
ICDHyr												
ICL												
LDH_D												
MALS												
MDH												
ME1												
ME2												
NADH11												
O2t												
PDH												
PFK												
PFL												
PGI												
PGK												
PGL												
PGM												
PIt												
PPC												
PPCK												
PPS												
PTAr												
PYK												
PYRt2r												
RPE												
RPI												
SUCCt2_2												
SUCCt2b												
SUCD1i												
SUCD4												
SUCOAS												
TALA												
AKGDH												
NADTRHD												
THD2												
TKT1												
TKT2												
TPI												

Figure 15.13: The effect of removing reactions from the core *E. coli* metabolic network on its ability to produce the biosynthetic precursors. The objective function is set to produce a precursor and the optimization is repeated for a series of cases where the flux has been set to zero for the reactions being studied. This procedure is performed for all the precursor molecules of interest. White areas indicate that the reaction deleted has no impact on production, gray areas indicate inhibited production, and black areas indicate that no production is possible. Prepared by Scott Becker.

The entries in Figure 15.13 can be studied:

- The knockout of the GLCpts (i.e., PTS) prevents the network from producing any of the precursors. The core model has no other way to import glucose. Similarly, if the phosphate transporter (PIt) is removed, none of the precursors that contain phosphate can be made.
- GAPD, PGK, and PGM, all in lower glycolysis, prevent the production of all the precursors except those that are found in the pentose pathway and in upper glycolysis. Without these enzymes there is no way that the core network can get carbon into the TCA cycle.
- The removal of three TCA cycle reactions (ACONT, CS, ICDyr) and of transhydrogenase (NADTRHD) and enolase (ENO) prevents the production of αKG.

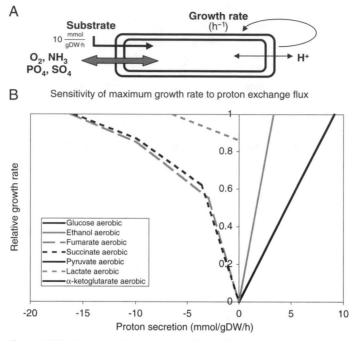

Figure 15.14: The growth rate as a function of the proton exchange rate with the environment. The core *E. coli* metabolic network was allowed to take up 10 mmol/gDW/hr of a carbon source and freely take up or secrete O_2, NH_3, PO_4, SO_4, and H^+ (A). A maximum growth rate could then be calculated as a function of the proton exchange rate (B). Prepared by Jennie Reed.

- The removal of three reactions in the pentose pathway (RPI, TALA, TKT1) prevents the production of R5P.

Almost all of these predictions of lethality are consistent with experimental data [51].

Similarly, the reduction in yield of the precursors is shown in the gray areas in the Figure 15.13. This table could be a map to experimentation, by supplementing the medium with the precursors that the knock-out strain cannot produce to determine if growth is then possible.

Effects of proton balancing

Metabolizing a substrate to support growth often requires the secretion or uptake of a proton. Thus, if a cell grows in an unbuffered environment, the pH of the medium can change as a result of proton exchange with the cellular surroundings. The core *E. coli* model was allowed to produce biomass from a variety of substrates as the limiting nutrient (Figure 15.14A). The optimal growth rate was then computed. In each case considered, the network needed to exchange a proton with the environment to achieve maximal growth. In some cases the import of a proton was required (making the environment alkaline), and in some cases the secretion of a proton was required (leading to an acidification of the growth medium). Carbon sources such as glucose will lower the medium pH while growth on other carbon sources (such as fumarate or pyruvate) will raise the pH of the medium.

The importance of the proton balancing to cell growth can be assessed by restricting the proton exchange rate with the environment. The growth rate can be computed as a function of the proton exchange rate (Figure 15.14B). In all cases the growth rate was a strong function of the proton exchange rate. For three substrates (pyruvate, lactate, α-ketoglutarate) reduced growth was achieved without any proton exchange, while for other substrates (glucose, ethanol, fumarate and succinate) no growth was possible without exchanging a proton. This example illustrates the importance of charge and elemental balancing of the reactions in a network.

15.6 Summary

➤ The fundamental subspaces of **S** are bounded with the application of v_{max} values.

➤ Specific points within these bounded solution spaces can be determined through optimization procedures.

➤ The optimization is based on a stated objective.

➤ Objectives can be used to probe network capabilities, to represent likely physiological objectives, and to represent candidate biological designs.

➤ If the objective function is linear, then linear programming can be used to find the optimal solution.

➤ Unique optimal solutions are found in the corners of the bounded solution space.

➤ Frequently, for large biological systems, the solutions are found on an edge of a surface of the solution space leading to redundant solutions. In such cases, many different solutions lead to the same optimal objective value.

➤ The computation of extreme pathways leads to the identification of the edges of the solution space. If the optimal solution lies on this edge, then the corresponding extreme pathway is the optimal solution.

➤ Network capabilities can be assessed and various physiological considerations can be addressed using LP.

15.7 Further Reading

Bonarius, H.P.J., Schmid, G., and Tramper, J., "Flux analysis of underdetermined metabolic networks: The quest for the missing constraints," *Trends Biotechnol*, **15**:308–314 (1997) .

Edwards, J.S., Covert, M.W., and Palsson, B.O., "Metabolic modelling of microbes: The flux-balance approach," *Environmental Microbiology*, **4**: 133–140 (2002).

Edwards, J.S., and Kauffman, K.J., "Biochemical engineering in the 21st century," *Current Opinion in Biotechnology*, **14**:451–453 (2003).

Edwards, J.S., Ramakrishna, R., Schilling, C.H., and Palsson, B.O., "Metabolic flux balance analysis," In S.Y. Lee and E.T. Papoutsakis (eds.), METABOLIC ENGINEERING, Marcel Dekker, New York (1999).

Nielsen, J., "It is all about metabolic fluxes," *Journal of Bacteriology*, **185**:7053–7067 (2003).

Price, N.D., Papin, J.A., Schilling, C.H., and Palsson, B.O., "Genome-scale microbial *in silico* models: The constraints based approach," *Trends in Biotechnology*, **21**:162–169 (2002).

Reed, J., and Palsson, B.O., "Thirteen years of building constraints-based models of *Escherichia coli*," *Journal of Bacteriology*, **185**:2692–2699 (2003).

Varma, A., and Palsson, B.O., "Metabolic flux balancing: Basic concepts, scientific and practical use," *Bio/Technology,* **12**:994–998 (1994).

Vasek, C., LINEAR PROGRAMMING, W.H. Freeman and Company, San Francisco, CA (1983).

CHAPTER 16

Parametric Sensitivity

One optimal solution is rarely of interest. The constraint-based optimization method introduced in the last chapter is scalable and can be repeatedly applied for varying environmental and genetic parameters. This scalability has spurred a growing number of analysis methods that have been developed under the constraint-based approach [176]. Here we describe some of the methods that are used to characterize a changing environment and genetic makeup. To date, the focus has been on the steady-state flux distributions, but now this approach is being used to study all allowable concentration [9] and kinetic states [55] and to algorithmatize iterative model-building procedures [33, 86].

16.1 Overview of Constraint-Based Methods

Constraint-based reconstruction and analysis (COBRA) procedures for analyzing the allowable phenotypic states of microorganism on a genome scale have developed rapidly in recent years [176, 183]. COBRA consists of three fundamental steps (Figure 16.1):

- First, as detailed in Chapter 3, a *genome-scale network reconstruction* (GENRE) is formed,
- Second, the appropriate constraints are applied to form the corresponding *genome-scale model* (GEM) *in silico*, and
- Third, various analysis methods are applied to evaluate the properties of GEMs.

This chapter focuses on the methods used in the third step.

It is important to note that only a curated and quality-controlled reconstruction can lead to organism-specific GEMs. Many automated procedures generate reaction maps that cannot be used as a basis for computation. The

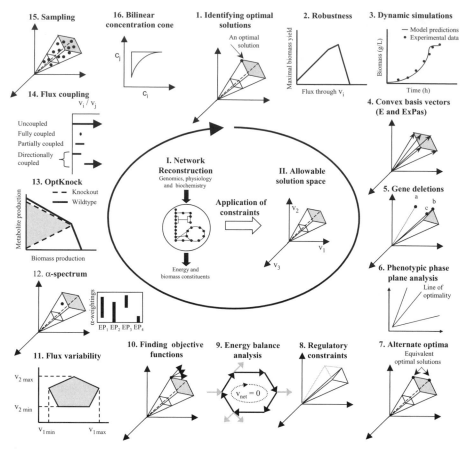

Figure 16.1: An overview of constraint-based methods used to analyze properties of reconstructed networks. Modified from [176].

reasons for this vary, but often basic rules of chemistry may be violated, pathways may have gaps, and so forth. The conversion of a GENRE to a GEM is a laborious and detailed process, but it has been accomplished for a number of organisms (Chapter 3).

Many *in silico* methods have been developed under the COBRA framework as outlined in Figure 16.1. These growing number of methods can be broadly classified into several categories:

1. Finding "best" or optimal states within the allowable range and studying the properties of such solutions. The methods in this category involve the computations of a single optimal solution, and the study of alternate, or equivalent, optimal solutions. One can also study the flux variability amongst all equivalent solutions.
2. Investigating how values for fluxes vary by changing one, two, or more parameters, and to investigate how strongly reactions are

coupled (correlated) in all functional states of networks. Often, the influence of environmental parameters can be varied over a specified numerical range.

3. Evaluating possible phenotypic changes as a consequence of genetic variations through the removal of one or more gene products from the network. These procedures have developed toward genome-scale designs of metabolic networks.

4. Defining and imposing additional constraints.

Optimization methods

The COBRA methods developed [176], some of which are discussed in this chapter, rely on the use of various optimization methods. These methods include:

- *Linear programming (LP):* this method is used when the problem to be solved involves a linear set of constraints (equalities and inequalities) and a linear objective function. LP was discussed in the last chapter and is used for flux-balance analysis (FBA).

- *Quadratic programming (QP):* this method is used when the problem to be solved involves a linear set of constraints (equalities and in-equalities) and a quadratic objective function. A quadratic objective arises when one uses a Euclidean distance as an objective function. When computing the Euclidean distance, the elements of a vector are raised to a second power. The "least-squares method" is an example of QP.

- *Mixed integer linear programming (MILP):* The formulation of LP problems often leads to the use of discontinuous variables. Often logical variables are introduced that take on a value of 0 or 1. MILP is used to solve this type of problem.

- *Nonlinear programming (NLP):* The most complicated optimization problems involve the use of nonlinear constraints and/or a nonlinear objective function. In general, such problems are hard to solve. One fundamental issue that arises is that the solution space being searched is nonconvex. In such a circumstance, one cannot guarantee finding the global optimum for the objective function in the space.

These optimization methods have been deployed in the various analysis methods developed under the COBRA umbrella. None of these methods are described in mathematical or algorithmic detail in this text. For such information one should consult established textbooks in the field such as [30, 8].

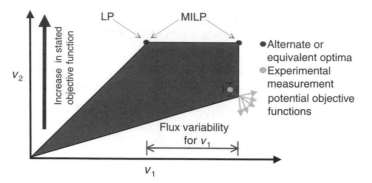

Figure 16.2: Determining optimal network states. If an objective function Z is stated to determine a functional state of a reconstructed network, the optimal solutions are found using optimization methods. LP will find one particular optimal solution. A MILP approach can be used to find all the alternate optimal solutions. Flux variability analysis can be used to find ranges of values for all the fluxes in the set of alternative optima. In this illustration, v_1 is the only flux variable between the alternate optima, whereas v_2 has the same numerical value ($Z = v_2$). If an objective function is not known, experimental measurements can be used to identify potential objectives that are optimized under the experimentally determined conditions. Taken from [176].

16.2 Evaluating Optimal States

As outlined in the last chapter, LP can be used to find single optimal solutions. The problem that is solved is:

$$\text{maximize } Z = \mathbf{w} \cdot \mathbf{v}$$
$$\text{subject to } \mathbf{Sv} = \mathbf{0} \tag{16.1}$$
$$\text{and } v_{i,\min} \leq v_i \leq v_{i,\max}, \quad i = 1, \ldots, n$$

Here, $\mathbf{v}$ contains both the internal and exchange fluxes. The solution to this problem may not be unique, as illustrated in Figure 16.2. Various ways to characterize the range of optimal solutions have been developed.

Alternative equivalent optima

Alternate flux distributions that lead to equivalent optimal network states are a property of genome-scale networks; recall Figures 15.8 and 15.9. The number of such alternate equivalent optima varies depending on the size of the metabolic network, the chosen objective function, and the environmental conditions. In general, the larger and more interconnected the network, the higher the number that can be realized. A GEM can reproduce the same overall functional state of a network in many different ways. The existence of equivalent optimal states is related to the biological notion of *silent phenotypes*. This feature is a network property and represents a distinguishing feature of the *in silico* modeling of phenotypes.

In comparison, solutions sought in the physicochemical sciences are typically unique (recall Table 2.1).

The methods used to compute alternate optima involve MILP [119, 168]. The basic problem stated in equation 16.1 gives a single optimal solution. A MILP-based approach can be iteratively implemented by adding additional constraints (so-called *integer cut constraints*). At iteration J, we eliminate one nonzero flux variable by stating

$$\sum_{i \in NZ^k} y_i \geq 1 \tag{16.2}$$

where y_i is a binary variable (i.e., either $y_i = 0$ or $y_i = 1$) that is associated with each reaction. NZ refers to the set of nonzero fluxes in a solution and $J - 1$ is used to index the previous iterations. Equation 16.2 makes sure that at least one of the nonzero fluxes in the previous solution is zero. Then we introduce another binary variable w_i for every reaction to make sure that previous solutions are not revisited and eliminate one nonzero flux

$$\sum_{i \in NZ^J} w_i \leq |NZ^k| - 1, \quad \text{for } k = 1, 2, \ldots, J - 1 \tag{16.3}$$

where $|NZ^k|$ denotes the number of members of the NZ set. Finally, we introduce two constraints:

$$w_i + y_i \leq 1 \quad \text{and} \quad w_i v_{i,\min} \leq v_i \leq w_i v_{i,\max} \tag{16.4}$$

for all i states that lead to "if $y_i = 1$ then $v_i = 0$ else $v_{i,\min} \leq v_i \leq v_{i,\max}$."

The outcome of such calculations is the identification of extreme points where the value of the objective function is identical; see Figure 16.2. Any point on the boundary in between such extreme points is also an equivalent optimal solution. The sets of such points are called *interior solutions*. This approach has been applied to a genome-scale model of *Escherichia coli* to identify a large number of equivalent solutions under a variety of growth conditions [184].

Flux variability

For a given optimal state (Z_{opt}), one can find the maximum and minimum allowable flux that a particular reaction can have while still supporting that functional state of the network. This range is computed for each flux v_i of interest by solving two LP problems:

$$\begin{aligned} &\text{maximize or minimize } v_i \\ &\quad \text{subject to: } Z = Z_{\text{opt}} \\ &\quad\quad\quad \mathbf{Sv} = \mathbf{0} \\ &\quad \text{and } v_{i,\min} \leq v_i \leq v_{i,\max}, \quad \text{for } i = 1, \ldots, n \end{aligned} \tag{16.5}$$

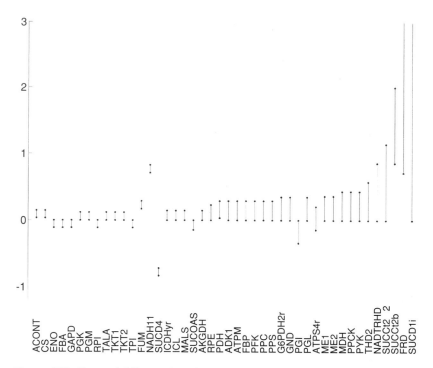

Figure 16.3: Flux variability analysis for the core *E. coli* model for optimal production of succinate from fumarate. Figure generated by Jennie Reed.

The first constraint fixes the value of the original objective function, the second constraint forces the flux balance to hold, and the third defines the allowable range for all the fluxes. To find the minimum value for v_i, the corresponding minimization problem is solved. This gives the range of the allowable numerical values for v_i as illustrated in Figure 16.2.

This procedure is similar to the one used to generate the α-spectrum in Chapter 13. The results can be graphically presented as the α-spectrum (Figure 13.20) with the individual fluxes on the x-axis and their allowable range on the y-axis.

Flux variability in the core *E. coli* model
Fixing Z at the optimal yield in Figure 15.8 and performing the procedure in equation 16.5 generates the range of fluxes allowable under optimal conditions (Figure 16.3). There are 88 extreme pathways that result in the same optimal yield.

Finding objective functions
As discussed in Chapter 15, the statement of an objective function has some inherent uncertainties and assumptions associated with it. If experimental data are available, the range of candidate objective functions that are

maximized under the measured conditions can be computed, as illustrated in Figure 16.2 [23]. The mathematics used are detailed and are beyond the scope of this text.

This procedure was applied to a network describing metabolism in *E. coli* where experimentally measured flux distributions were available [196]. The back-calculated objective functions for aerobic and anaerobic growth were similar to each other. The set of objective functions back-calculated were similar to biomass generation, thus indicating that one metabolic objective function can predict both aerobic and anaerobic flux distributions and the behavior is consistent with the maximization of biomass generation.

16.3 Varying Parameters

The sensitivity of the optimal properties of a network can be assessed by changing parameters over a given range of values and repeatedly computing the optimal state. Both environmental and genetic parameters can be considered.

Robustness analysis: varying one parameter

One parameter can be varied in a stepwise fashion and the LP problem solved for every incremental value. If we are interested in varying v_j between two values, i.e., a and b, we can solve

$$
\begin{aligned}
&\text{maximize } Z_k = \mathbf{w} \cdot \mathbf{v} \\
&\text{subject to } v_j = c_k \\
&\qquad\qquad \mathbf{Sv} = \mathbf{0} \\
&\text{and } v_{i,\min} \le v_i \le v_{i,\max} \quad i = 1, \dots, n, i \ne j
\end{aligned}
\tag{16.6}
$$

l times, where c_k is varied in l increments between a and b; i.e., from $c_1 = a$ to $c_l = b$ with $c_{k+1} = c_k + (b - a)/(l - 1)$. The results will generate a series of l values for Z ($Z_k, k \in [1, l]$), and the associated shadow prices and reduced costs.

Varying oxygen uptake rate in the core *E. coli* network

The effects of varying the oxygen uptake rate of the core *E. coli* metabolic network to support optimal growth can be computed using the procedure in equation 16.6. The effects of varying the oxygen uptake rate from zero (completely anaerobic growth) to all the oxygen required for fully oxidizing the substrate on the growth rate are shown in Figure 16.4.

As the oxygen uptake rate increases from zero, the growth rate increases (Figure 16.4). There are three linear segments in the rising part of the

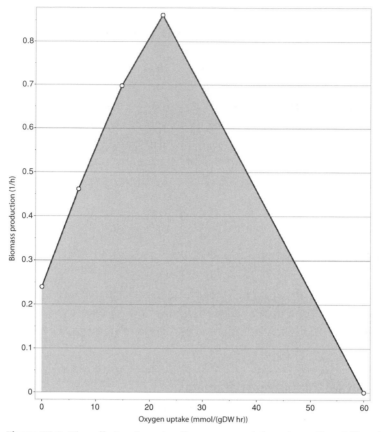

Figure 16.4: The effects of varying the oxygen uptake rate on the ability of the core *E. coli* network to support growth. The uptake rate of glucose is 10 mmol/gDW/h. The lines are the maximal growth rates, while the area represents all allowable biomass production rates (they are suboptimal growth states). The optimal point corresponds to the perfect conversion of glucose into biomass with no bioproduct formation. The shadow prices for fermentation bioproducts for oxygen uptake rate between 0 and 22 mmol/gDW/h are shown in Table 16.1. Prepared by Scott Becker.

curve. The shadow price structure changes at each discontinuity. In the first segment, the shadow prices for acetate, formate, and ethanol are zero (Table 16.1). These three metabolites are useless to the cell and are thus secreted. In the second segment, the shadow prices for acetate and formate are zero and are secreted, while ethanol has value to the growth process as indicated by the negative shadow price. In the third segment, only acetate has a zero shadow price and is secreted. At the peak of the curve, the *carbon to oxygen uptake* (C/O) ratio is perfect for biomass formation and no by-products are secreted. Beyond the peak in the curve, too much oxygen is taken up relative to glucose and the growth rate drops due to forced dissipation of the excess oxygen. This segment represents an unrealistic physiological situation as discussed below.

Table 16.1: The changes in the secretion rate and shadow prices of key metabolites in the core *E. coli* metabolic network with varying oxygen uptake rate. The number given is the secretion rate in mmol/gDW/h, followed by the shadow price in parentheses. Metabolites with a zero shadow price do not affect the value of the objective function (see equation 15.4) and are secreted. Metabolites with a negative shadow price can increase the objective function and are thus not secreted.

Oxygen uptake	Acetate	Formate	Ethanol	Lactate	Succinate
0	8.087 (0)	17.08 (0)	8.262 (0)	0 (−0.0045)	0 (−0.011)
6.66	12.988 (0)	14.933 (0)	0 (−0.0030)	0 (−0.0059)	0 (−0.012)
14.7	7.382 (0)	0 (−0.0034)	0 (−0.0095)	0 (−0.012)	0 (−0.022)
22.35	0 (−0.046)	0 (−0.011)	0 (−0.069)	0 (−0.069)	0 (−0.080)

Phenotypic phase planes: varying two parameters

Robustness analysis represents the effects of varying a single parameter on the performance of a network. In a similar fashion, two parameters can be varied simultaneously, Figure 16.5 [13, 48]. A set of variables that have

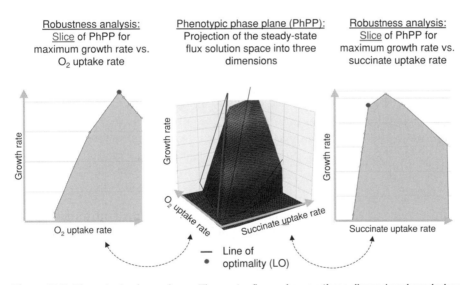

Figure 16.5: Phenotypic phase planes. The center figure shows a three-dimensional rendering of a maximal growth rate plotted as a function of two variables: the O_2 and succinate uptake rates. A phase plane is a projection of this surface into two dimensions, the floor of the 3D figure. The line of optimality corresponds to the conditions where the objective function is optimal (in this case, growth rate). Robustness analysis of the two uptake rates individually is shown in the two side panels. The graph on the left shows the effect on growth rate on varying O_2 uptake at a fixed succinate uptake rate (as in Figure 16.4) and represents a slice through the 3D surface at a specific oxygen uptake rate. Conversely, the graph on the right shows the effect on biomass generation on varying the succinate uptake rate at a fixed oxygen rate. From [176].

been of particular interest are the substrate and oxygen uptake rates for microbial growth. Then the optimal flux-maps can be calculated for all points in a plane formed by using the substrate uptake rate on the x-axis and the oxygen uptake rate on the y-axis. A 3D surface can be graphed above this x, y plane.

The plane formed by the two uptake rates is called the *phenotypic phase plane* (PhPP). One can denote each phase as $Pn_{x,y}$, where P represents phenotype, n is the number of the demarcated region for this phenotype, and x, y are the two uptake rates on the axes of the plane. The PhPP in some ways resembles the phase planes used in physical chemistry, which define the different states (i.e., liquid, gas, or solid) of a chemical system depending on the external conditions (e.g., temperature, pressure).

Although one can compute an infinite number of optimal solutions in the PhPP, it turns out that there are a finite number of fundamentally different optimal functional states present in the PhPP. The demarcations between the regions of different functional states can be determined from the shadow prices of the variables that are represented on the axes of the PhPP. As the robustness analysis shows, the shadow prices can be used to interpret shifts from one optimal flux distribution to another. This procedure leads to the definition of distinct regions, or phases, in the PhPP in which the optimal use of the network is fundamentally different, corresponding to a different functional state.

The regions in the PhPP can be defined based on the contributions of the two parameters represented on the x- and y-axes to the objective function. To facilitate such an interpretation, we define the ratio of the relative shadow prices for the two variables on the axes of the PhPP:

$$\alpha = -\frac{\pi_x}{\pi_y} \tag{16.7}$$

where π is the shadow price (see equation 15.4) and x and y refer to the variables on the x- and y-axes. The negative sign on α is introduced in anticipation of its interpretation. The ratio α is the relative change in the objective function for changes in the two key exchange fluxes. In order for the objective function to remain constant, an increase in one of the exchange fluxes will be accompanied by a decrease in the other, and thus we introduce the negative sign on the definition of α. The parameter α is thus the slope of a line in the PhPP along which the value of the objective function is a constant. This line is called an *isocline*.

The slope of the isoclines within each phase of the PhPP is calculated from the shadow prices. Thus, the slope of the isoclines will be different in each region of the PhPP. Based on these considerations, we identify four

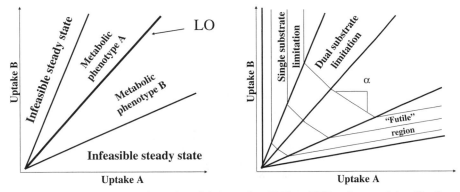

Figure 16.6: Characteristics of the PhPP. (A) Phases in a PhPP and (B) isoclines and classification of different phases.

types of regions on the PhPP (Figure 16.6):

1. In phases where the α value is negative, there is dual limitation of the substrates. Based on the absolute value of α, the substrate with a greater contribution toward obtaining the objective can be identified. If the absolute value of α is greater than unity, the substrate along the x-axis is more valuable toward obtaining the objective, whereas if the absolute value of α is less than unity, the substrate along the y-axis is more valuable to the objective.

2. The phases where the isoclines are either horizontal or vertical are phases of single substrate limitation; the α value in these phases will be zero or infinite, respectively. These phases arise when the shadow price for one of the substrates goes to zero, and thus has no value to the cell.

3. Phases in the PhPP can also have a positive α value; these are termed "futile" phases. In these phases, one of the substrates is inhibitory toward obtaining the objective function, and this substrate will have a positive shadow price. The metabolic operation in this phase is wasteful, in that it consumes substrate that is not needed to improve the objective, i.e., the post-peak segment in Figure 16.4. Phases with positive α values are expected to be physiologically unstable. For example, under selection pressure, cells would move their phenotype state out of the phase.

4. Finally, due to stoichiometric limitations, there are infeasible steady-state phases in the PhPP. If the substrates are taken up at the rates represented by these points, the metabolic network is not able to obey the mass, energy, and redox constraints while generating biomass. The metabolic network can only transiently operate in such a region.

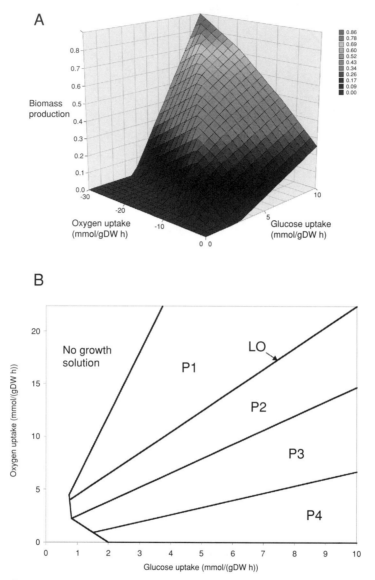

Figure 16.7: The phenotypic phase plane for the core *E. coli* model for growth on glucose. (A) A 3D surface of maximal biomass production as a function of glucose fixed oxygen and glucose uptake rates. (B) The PhPP, which is the "floor" of the polytope in (A). The lines are projections of the edges of the polytope in (A). Prepared by Scott Becker.

PhPP for the core *E. coli* model

The phase plane for aerobic growth on glucose is shown in Figure 16.7. The PhPP (Figure 16.7B) has four phases. They are consistent with the robustness analysis in Figure 16.4. At low oxygen uptake rates, the phase $P4_{O_2,gluc}$ is characterized by the secretion of acetate, formate, and ethanol. $P3_{O_2,gluc}$ has acetate and formate secreted, and $P2_{O_2,gluc}$ has only acetate secretion. These three phases have isoclines with negative slopes. The line of optimality (LO) represents the condition where all the carbon taken up is fully oxidized or incorporated into biomass. $P1_{O_2,gluc}$ has isoclines with a positive slope and are thus phenotypically unstable, i.e., if the glucose rate is fixed, then lowering the oxygen uptake rate toward the LO will increase the growth rate. Thus, the maximum allowable oxygen uptake rate would not be chosen to maximize growth rate.

16.4 Changing the Genotype

Many situations of interest involve the analysis of the consequences of the loss or gain of gene function. These may result from the removal of one or more genes from a genome that results in the loss of a reaction in a network. One may also be interested in the design of new strains and determine what genes need to be added or removed in order to design a strain that can achieve certain functional states of interest. The gene deletion analysis shown in Figure 15.13 and the computations on the lethal lines in Figure 7.5 are examples of analyzing the consequences of gene deletions. Several sophisticated *in silico* analysis methods have been developed to carry out analyses of the consequences of gene deletions. They are graphically illustrated in Figure 16.8 and will be briefly discussed in this section.

Minimization of metabolic adjustment (MOMA)

This method was developed to predict the changes in the location of the flux vector within the solution space if the function of a gene product is lost [203]. A loss of gene product may reduce the solution space. If an optimal functional state for the wild-type strain was in the portion of the solution space that is eliminated with the loss of the function of a gene product, it will have to be projected into the reduced solution space to represent functions of the knockout strain. MOMA finds a new solution in the reduced solution space such that the Euclidean distance between the wild-type state and the reduced solution space is minimized. Since the Euclidean distance is not a linear function, this procedure uses quadratic programming.

Mathematically the problem is stated as:

$$
\begin{aligned}
&\text{Minimize } Z = (\mathbf{v}_{KO} - \mathbf{v}_{wt})^{\mathrm{T}} \, (\mathbf{v}_{KO} - \mathbf{v}_{wt}) \\
&\text{subject to } \mathbf{Sv} = \mathbf{0} \\
&\quad \text{and } v_j = 0 \text{ resulting from the knockout} \\
&\quad \text{except } v_{i,\min} \leq v_i \leq v_{i,\max}, \quad i = 1, \dots, n \quad i \neq j
\end{aligned}
\tag{16.8}
$$

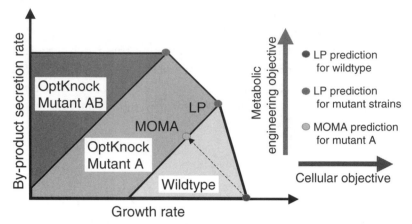

Figure 16.8: Illustration of an altered solution space resulting from gene knockout and *in silico* methods for the analysis of the consequences of a gene knockout. A wildtype solution space and two smaller knockout solution spaces are illustrated. Optimization of growth rate (x-axis) in the wildtype (square) would not produce a metabolic by-product (y-axis). Optimization of growth rate in the two mutant strains (triangles) finds solutions with by-product secretion. Minimization of metabolic adjustment (MOMA) minimizes the difference between the optimal metabolic state of the wild type and the reduced space (circle). OptKnock can be used to identify knockouts strains that couple optimal biomass production with by-product secretion. OptKnock identifies gene knockouts such that in order for a cell to grow optimally, it must produce the desired by-product (OptKnock Mutants A and AB – triangles). Figure adapted from [176].

where $\mathbf{v}_{KO}$ is the flux distribution in the knockout strain, $\mathbf{v}_{wt}$ is the optimal solution for the wild-type strain, and the index j represents the fluxes that are removed from the network as a result of the gene knockout.

MOMA analysis of core *E. coli*

The MOMA projection of the optimal growth state of the core *E. coli* model following the deletion of PFK corresponds to a biomass of 0.0108 g/mol glucose. This point in the reduced solution space is the closest (using Euclidean distance) to the optimal biomass production of 0.0861 g/mol with the full set of reactions. However, an LP search (equation 16.1) over the reduced space finds an optimal yield of 0.0704 g/mol. The corresponding flux distributions are shown in Figure 16.9. The LP search identifies a solution that is essentially a textbook use of the classical pathway, with the additional use of PPC as an entry route to the TCA (Figure 16.9A). The LP optimization of the PFK knockout produces a flux map where glucose is processed through the pentose pathway to generate NADPH. The transhydrogenase converts NADPH to NADH that then donates the electrons to the ETS to produce ATP. The TCA is thus partially used. The MOMA projection changes this pathway significantly (Figure 16.9B). In the MOMA solution, a PPC-PPCK cycle appears, the glyoxylate shunt and transhydrogenase is

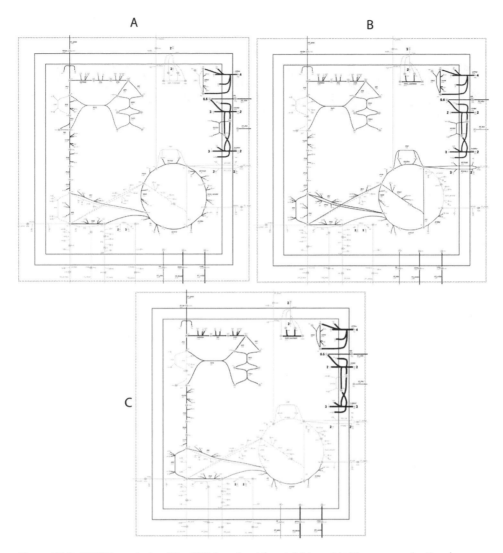

Figure 16.9: MOMA analysis of the PFK knockout for stoichiometric biomass production (no maintenance). (A) The wild-type flux distribution (yield = 0.0861 g/mol). (B) The MOMA projection (yield = 0.0108 g/mol). (C) The LP solution for the PFK knockout (yield = 0.0704 g/mol). Prepared by Scott Becker.

used, and quite a bit of succinate is produced. This flux map might be taken as a prediction of the initial response to the loss of PFK.

Bilevel optimization procedures

Bilevel optimization is the nesting of objectives. In this way, one can synchronize physiological and engineering objectives [25, 169, 170]. One can put an outer optimization problem over the *E. coli* biomass optimization problem. This outer problem is used to find the minimum number of gene

Table 16.2: The minimum gene deletion sets in the core *E. coli* network that lead to succinate secretion under optimal growth conditions. Prepared by Markus Herrgard.

Strain	Reactions eliminated	Growth rate (1/h)	Succinate secretion (mmol/gDW/h)	Oxygen uptake (mmol/gDW/h)
WT	0	0.86	0.00	22.35
CyoA-/Pyka-/Pykf-	2	0.16	6.06	0.00
CyoA-/Pflc-	2	0.21	0.78	0.00
AceE-/Sdh-/Zwf-	3	0.48	6.34	12.42
AckA-/FumA-/FumB-/FumC-/Zwf-	3	0.58	5.61	15.11
Pgi-/Pta-/Sdh-	3	0.69	3.14	19.01
Ppsa-/Pta-/Sdh-	3	0.72	2.24	20.68

knockouts or gene additions that maximizes the secretion of a by-product of interest while the inner problem maximizes the biomass formation. The bilevel optimization procedure has been used to compute optimal designs based on the genome-scale *E. coli* model. Some of these designs have been implemented for the production of lactate in *E. coli* [62].

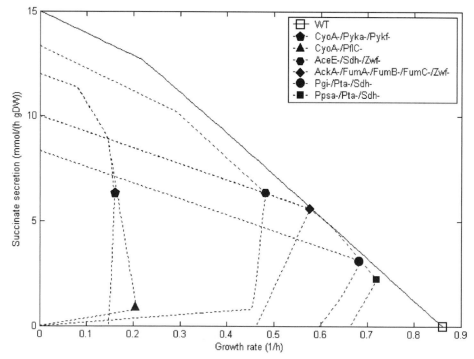

Figure 16.10: The succinate secretion rate graphed as a function of the growth rate for the core *E. coli* network, and the deletion sets given in Table 16.2. The maximum growth rates are indicated by the points as shown. Prepared by Markus Herrgard.

Analysis of the core *E. coli* network

The bilevel optimization procedure can be used to calculate the gene deletion sets that lead to the secretion of succinate as growth rate is optimized. The five smallest gene deletion sets are shown in Table 16.2. The relationship between the growth rate and the succinate secretion rate is shown in Figure 16.10. For the core network, this curve decreases monotonically. The maximal growth rate is found at the point represented by the open square where there is no succinate secreted. The bilevel optimization computed deletion sets give a curve that has a maximal growth rate at a finite secretion rate of succinate (solid points in Figure 16.10). To optimize a bioprocess, a balance between growth and secretion rate is required [234]. The best balance tends to favor a small drop in the growth rate, relative maximal growth, and a small per-cell secretion rate of the desired product.

16.5 Summary

> Constraint-based reconstruction and analysis at the genome scale has proved to be useful.

> A growing number of methods to analyze the properties of genome-scale networks have emerged. These methods are principally based on the use of constraint-based optimization.

> Various analysis methods that look at optimal states, parameter variations, and the consequences of gene knockouts have been developed.

16.6 Further Reading

Kauffman, K.J., Prakash, P., and Edwards, J.S.,"Advances in flux balance analysis," *Current Opinion in Biotechnology*, **14**:491–496 (2003).

Price, N.D., Reed, J.L., and Palsson, B.O., "Genome-scale models of microbial cells: Evaluating the consequences of constraints," *Nature Reviews Microbiology*, **2**:886–897 (2004).

Reed, J.L., and Palsson, B.O., "Thirteen years of building constraints-based *in silico* models of *Escherichia coli*," *Journal of Bacteriology*, **185**:2692–2699 (2003).

CHAPTER 17

Epilogue

Now that we have come to the end of this text, it is time to ponder what we have done, how far we have come, and what lies ahead. In this chapter I put forth some of my thoughts related to these issues.

17.1 Types of questions asked in biology

There are fundamentally three types of questions that are asked in biology: "what," "how," and "why."

What is there?

We have made substantial strides in answering this type of question. We can sequence entire genomes and use bioinformatic analyses to determine what is in a genome. We can expression profile a genome under various conditions. We now have extensive information about genomes, cells, and organisms, and are in a position to continue to generate much more. It is indeed this impressive availability of data that has made biology "data-rich" and has been the driving force for the emergence of systems biology.

How does it work?

Science seeks to generate mechanisms and theories to explain the world around us. Functional genomics tries to assign function to various gene products and segments of a genome. The large number of interactions that needs to be taken into account to explain cellular components has grown substantially with our growing knowledge of cellular components. The drive to reconstruct genome-scale networks and to assess their functional states is a response to this need. This book focused on biochemically, genetically, and genomically based "bottom-up" approaches to answer the "how" questions systematically and at a network level.

Why does it work the way it does?

The answers to such questions are generally very difficult to obtain. In biology, they are based on the understanding of evolution and making teleological arguments. This book does not address such issues, although the statement of biological objectives and assessing their importance borders on such questions. Surprisingly, the outcome of adaptive evolution in bacteria can be predicted in genome-scale models with about 70%–80% success rate [62, 64, 99].

17.2 Why make models?

Mathematical modeling has been practiced in various branches of science and engineering. The purpose and utility of model building has been succinctly summarized and discussed [5]; quoting:

1. To organize disparate information into a coherent whole;
2. To think (and calculate) logically about what components and interactions are important in a complex system;
3. To discover new strategies;
4. To make important corrections to the conventional wisdom; and
5. To understand the essential qualitative features.

All of these issues were directly or indirectly addressed in this book.

17.3 Expanding the scope

The scope of material in this book is likely to grow in two categories: biological content and the range of *in silico* analysis methods covered.

Genome-scale models should be useful for addressing all the five issues listed above in that they relate the contents of genomes to their respective living processes. In a sense, genome-scale models "bring genomes to life." Most of the material in this book is related to metabolism and microorganisms. This scope is likely to change if there are subsequent editions of this book. We are bordering on having high-resolution reconstructions of signaling and transcriptional regulatory networks that will enable us to use the methods described herein to analyze their properties. With 99% of the human euchromatin sequence finished [32], we are now in a position to reconstruct the human metabolic map. Once that is accomplished, the materials and methods described in this book will hopefully become useful to study human physiology and pathophysiology.

There is considerable interest in characterizing the dynamic states of reconstructed networks. Here we have primarily dealt with the steady state, but interestingly, in nonlinear dynamical analysis, the properties of the steady state are the most informative with respect to the dynamic characteristics of the network. A full treatment of network dynamics is a highly mathematical subject and is not included in this text. There is clearly room for an additional text on dynamic analysis methods of large biochemical reaction networks. However, such an undertaking may have to wait a bit until we have more information available about the numerical values of the kinetic constants that are needed for dynamic analysis. Ideally, the grand challenge of predicting kinetic properties from DNA sequences would offer an efficient solution to this issue. Another topic on *in silico* analysis that needs to find its way in to a textbook of this type is the various statistical procedures that are used for top-down analysis of networks. These methods will mature and solidify in the coming years.

17.4 Where does the field need to go?

Systems and network analysis in biology is at an early stage of development. We have made some progress, but there is a long way to go. There are a few challenges that the field faces, some are generic and some specific. A few of these challenges are as follows:

Dealing with physical vs. biological causation: as illustrated in Figure 12.3 these two issues are at opposite ends of a hierarchy of events that we have to deal with. It is clear that this relationship is hierarchical and calls for multiscale analysis.

Consequently, one challenge is to determine what the information content is in the various omics data types since they address different layers in the hierarchy. Constraints at the lowest level must hold at all higher levels. However, there will be additional constraints and considerations that arise as we move up the hierarchy. Thus, there may be measurable changes at a lower level that are inconsequential at a higher level. The existence of hitchhiker mutations is one example, and we can expect to find similar examples with other omics data types. *In silico* analysis methods are needed that explicitly deal with these issues.

In the intracellular environment there is much *thermal noise*, as biological components are constantly bouncing into one another. The question of "who talks to whom?" that has been getting much attention, particularly in the study of signaling networks, should perhaps more clearly be asked as "who listens to whom?" since biological components can be bouncing into one another without any resulting chemical reaction or any influence on the network as a whole. Reactions that lead to no further consequences would be "dead ends" in a network. Dead ends can turn into "live contacts" if new links are established. Robust and properly functioning networks within the constraints of thermal noise are clearly important in maintaining basic cellular functions.

A molecule within the complex intracellular environment (see Figure 2.3) will, by virtue of physicochemical constraints, only be able to interact with a finite number of potential partners. Thus, every molecule and every component seems to function in what might be called a *small world*. What is happening in one small world or in one locality inside a cell may be unknown to what is happening a few locales over, at least not at every point in time. Therefore, what is happening at one location inside a cell may be only loosely connected to what is occurring elsewhere. In terms of a computer programming language, this may be thought of as parallel processing, which every so often needs to be synchronized by some higher level organization.

This necessarily leads to the consideration of the 3D arrangements of cells and the morphogenic properties of groups, or modules, executing integrated functions in molecular biology. Therefore, the 2D representations of biochemical reaction networks that are discussed in Part I of this text will eventually have to become three dimensional. The consideration of the architecture of cellular processes at the ~100 nm length scale is likely to lead to an exciting new dimension in systems biology, in terms of both *in silico* analysis as well as new generation of measurement tools.

Hierarchical analysis will come down to aggregating or combining the elementary variables (e.g., concentrations and fluxes) into new quantities that will systematically take us from chemistry to biology. Mathematical definitions of aggregate variables in terms of pools and pathways have appeared [186]. The systematic decomposition of pools appears through a combination of network topology and kinetic values through the use of temporal decomposition and modal analysis [109, 150]. Such analyses begin to focus on variables such as the capacity to carry a particular property (such as the adenosines to carry high energy bonds) and how occupied such capacities are [186]; recall Chapter 10. A systematic analysis of this sort shows for example that the ca. 40 variables that describe metabolism in the human red cell lead to the definition of four aggregate variables that correspond to the four key physiological functions of the red cell. The systemic reactions defined in Chapter 6 are another example of a high-level network property.

As we become better at biologically driven hierarchical analysis of large networks, we will begin to be able to formulate mathematical definitions of key properties like redundancy, robustness, causality, and so on. Potentially, such analyses may graduate to definitions of network properties that may relate to fundamental biological properties such as "what is self."

These considerations beg the question of how cells function, or "think" and "remember." It appears that on a short time scale, cells respond to their environment and "look up" functional states that have been learned and are stored in the cell's "memory." Memory here is a nebulous term, of course, but it should reflect a selected functional network state, the links used, and their kinetic properties. Mechanism is needed to "look up" this entry in the cell's memory, which would correspond to a signaling and regulatory network. Of course, over longer time scales, the contents of the memory can change. Presumably, a cell will only remember a finite number of network states that will be remembered based on recent survival needs. As new behaviors are learned and stored, older ones may have to disappear.[1]

17.5   Closing

Hopefully this text will be useful to those who are interested in network reconstruction, the biochemically and genomically accurate representation of such reconstructions, and methods to interrogate the functional states of networks. It is the author's intent to complement the book with homework sets and the continued posting of genome-scale networks for all to use (on http://systemsbiology.ucsd.edu). It is also clear that this book is just the first installment of a number of books in this growing field of systems biology, books that will grow in scope and educational impact.

[1] This item was inspired by Sydney Brenner.

Nomenclature and Abbreviations

Roman Symbols[1,2]

$\mathbf{A_v}$	the left adjacency matrix of $\mathbf{S}$ (equation 7.4)
$\mathbf{A_x}$	the right adjacency matrix of $\mathbf{S}$ (equation 7.7)
$\mathbf{a}$	a vector of conservation pool sizes (equation 10.6)
a_i	size of conservation pools, i, in units of concentration (Figure 10.3)
$\mathbf{B}$	matrix of spanning edges (equation 14.4)
$\mathbf{b}$	exchange flux vector (equation 15.1)
b_i	vectors representing spanning edges (equation 14.4); also individual exchange fluxes (equation 6.23)
$\mathbf{E}$	the elemental matrix represents the elemental composition of all the compounds considered in a network. The columns correspond to the compounds and the rows to the elements (equation 6.18).
$\mathbf{e}_i$	a row vector in the elemental matrix giving the elemental composition of compound i (equation 6.19)
F_i	cumulative singular value (equation 8.3)
f_i	fractional singular value (equation 8.2)
$\mathbf{I}$	the identity matrix (equation 8.4)
k_i	rate constant (equation 11.9)
$\mathbf{L}$	a matrix of left null space basis vectors (equation 10.1)
$\mathbf{l}_i$	a left null space basis vector, a row vector (equation 10.3)
$\mathbf{P}$	pathway matrix (Figure 9.10)

---

[1] Equation number given indicates the first appearance of symbol in the text.
[2] Some symbols are used to designate more than one quantity. Standard nomenclature used in the literature dictates such dual use.

$\mathbf{P}_{\mathrm{LM}}$ — pathway length matrix (equation 13.5)

$\overline{p}$ — expected value (equation 14.11)

$\mathbf{p}$ — the number of extreme pathways (equation 9.11)

$\mathbf{p}_i$ — pathway vector (equation 9.7)

$\mathbf{R}$ — a matrix of right null space basis vectors (equation 9.2)

$\mathbf{R}_{\mathrm{PM}}$ — reaction participation matrix (equation 13.6)

$\mathbf{r}_i$ — a right null space basis vector, a column vector (equation 9.3); also row in the stoichiometric matrix (equation 11.11)

$\mathbf{S}$ — the stoichiometric matrix; each row corresponds to a metabolite and every column to a reaction. Dimension given in terms of $n x m$ (equation 6.3).

$\mathbf{s}_i$ — the reaction vector corresponding to reaction i, a column vector of the stoichiometric matrix (equation 6.7)

s_{ij} — the ijth element in the stoichiometric matrix (equation 6.5)

$\mathbf{U}$: — matrix of left singular vectors (equation 8.1)

$\mathbf{V}$: — matrix of right singular vectors (equation 8.1)

$\mathbf{v}$ — the vector of reaction fluxes, dimension is n (equation 6.1)

v_i — the flux through the ith reaction; units are moles per volume per time (equation 6.1).

$\mathbf{w}$ — the vector of weights (equation 15.2)

w_i — element i in a vector of weights (equation 9.3)

$\mathbf{x}$ — the vector of concentrations, dimension is m (equation 6.2)

$\mathbf{x}'$ — the time derivative of $\mathbf{x}$ (equation 6.4)

$\dot{\mathbf{x}}$ — the time derivative of $\mathbf{x}$ (equation 6.4)

x_i — the concentration of the ith compound; units are moles per volume (equation 6.2)

Z — an objective function, a scalar (equation 15.2)

Greek Symbols

α — slope of an isocline (equation 16.7)

α_k — weights on pathway vectors (equation 9.8)

Γ — mass action ratio (equation 14.14)

ϵ — estimated error (equation 14.12)

μ — average value (equation 14.12)

ξ — vector of ξ_i (equation 10.10)

ξ_i — used for parameterization ($\xi_i \in [0, 1]$) (equation 10.10)

π_i	participation number (equation 7.2); also shadow price (equation 15.4)
ρ_i	connectivity number (equation 7.3); also reduced cost (equation 15.5)
Σ	diagonal matrix of singular values (equation 8.1)
σ_i	singular value (equation 8.2); also standard deviation (equation 14.11)

Mathematical Symbols

Col(A)	column space of matrix A
dim($\cdot$)	dimension of a vector, space, matrix, ...
Left Null(A)	left-null space of matrix A
Null(A)	null space of matrix A
Row(A)	row space of matrix A
T	transpose
$\|\cdot\|$	the norm of a vector
($\equiv$)	row vectors in a matrix
($\|\|\|$)	column vectors in a matrix

Subscripts

dyn	dynamic
exch	exchange
int	internal
ref	reference
ss	steady state
tot	total

Superscripts

$\wedge$	binary form (equation 7.1)
$-$	expected value (equation 14.11)

Abbreviations

3PG	3-phospho-D-glycerate
αKG	Alpha ketoglutarate
AcCoA	Acetyl coenzyme A
ADP	Adenosine diphosphate

ADK	Adenylate kinase
AKGDH	Alpha ketoglutarate dehydrogenase
AMP	Adenosine monophosphate
ArcA	Aerobic respiration control
Arg	Arginine
ATP	Adenosine triphosphate
ATPM	ATP maintenance requirement
ATPS4r	ATP synthesis (four proteins for one ATP)
BIGG	Biochemically, genetically, and genomically structured data base
C/O	Carbon to oxygen ratio
CAMP	Cyclic AMP
CBP	CREB binding protein
ChIP	Chromatin immunoprecipitation
CMR	Comprehensive microbial resource
COBRA	Constraint-based reconstruction and analysis
CoSets	Correlated sets of reactions
CREB	CAMP response element binding protein
CRP	C-reactive protein precursor
CYGD	Comprehensive yeast genome database
CYTBD	Cytochrome oxidase bd
DHAP	Dihydroxyacetone phosphate
DNA	Deoxyribonucleic acid
DOQCS	Database of quantitative cellular signaling
E4P	D-erythrose 4-phosphate
E.C.	Enzyme commission
EcoCyc	*Escherichia coli* encyclopedia
EGF	Epidermal growth factor
ENO	Enolase
ETS	Electron transfer system
ExPa	Extreme pathway
F6P	D-fructose 6-phosphate
$FADH_2$	Flavin adenine dinucleotide – reduced
FBA	Flux-balance analysis
FBP	Fructose-biphosphatase
FCF	Flux coupling finder

FDP	D-fructose 1,6-biphosphate
FMNH$_2$	Flavin mononucleotide – reduced
FNR	Fumarate and nitrate reduction
FRD	Fumarate reductase
FRET	Fluorescence resonance energy transfer
G3P/GAP	Glyceraldehyde 3-phosphate
G6P	D-glucose 6-phosphate
G6PDH	Glucose-6-phosphate dehydrogenase
GAL	Galactose
GAPD	Glyceraldehyde 3-phosphate dehydrogenase
GDP	Guanosine 5-diphosphate
GEM	Genome-scale model
GENRE	Genome-scale network reconstruction
GFP	Green fluorescent protein
GLIMMER	Gene locator and interpolated markov modeler
GLU	Glucose
GND	Phosphogluconate dehydrogenase
GOLD	Genomes on-line database
GPCR	G-protein-coupled receptors
GPR	Gene-protein reaction
GRN	Genetic regulatory network
GTP	Guanosine triphosphate
GWLA	Genome-wide location analysis
GyrA	DNA gyrase
HOG	High osmolarity glycerol
HT	High-throughput
I/O	Input-to-output
ICAT	Isotope-coded affinity tags
IKK complex	I kappa β kinase
IOFA	Input/output feasibility array
JAK	Janus-associated kinase
KEGG	Kyoto encyclopedia of genes and genomes
LigA	DNA ligases
LIPID MAPS	Lipid metabolites and pathways strategy
LO	Line of optimality
LP	Linear programming

MAPK	Mitogen-activated protein kinase
McrA	Modified cytosine restriction
MILP	Mixed integer linear programming
MOMA	Minimization of metabolic adjustment
mRNA	Messenger ribonucleic acid
NAD	Nicotinamide adenine dinucleotide
NADH	Nicotinamide adenine dinucleotide – reduced
$NADP^+$	Nicotinamide adenine dinucleotide phosphate
NADPH	Nicotinamide adenine dinucleotide phosphate – reduced
NADTRHD	Nicotinamide adenine dinucleotide phosphate transhydrogenase
NLP	Nonlinear programming
OAA	Oxaloacetate
OmpH	Outer membrane protein
ORF	Open reading frame
P/O	Phosphate-to-oxygen
PDH	Pyruvate dehydrogenase
PEP	Phosphoenolpyruvate
PFK	Phosphofructokinase
PGI	Glucose 6-phosphate isomerase
PGK	Phosphoglycerate kinase
PGL	6-Phosphogluconolactonase
PGM	Phosphoglycerate mutase
pH	Potential of hydrogen
PhPP	Phenotypic phase plane
PIT	Phosphate transporter
PK	Pyruvate kinase
PKA	Protein kinase A
pKa	Ionization constant
Po1A	DNA polymerases
PrbAMP	Phosphoribosyl AMP
PTS	Proton transfer system
PYK	Pyruvate kinase
PYR	Pyruvate
QP	Quadratic programming
R5P	Alpha-D-ribose 5-phosphate

RecA	Renaturation protein
RegulonDB	Regulon database
RNA	Ribonucleic acid
ROB	Right origin-binding protein
ROOM	Regulatory off–on Minimization
RPE	Ribulose 5-phosphate 3-epimerase
RPI	Ribose 5-phosphate isomerase
RpoA/D	RNA polymerase
SCPD	*Saccharomyces cerevisiae* promoter database
SGD	*Saccharomyces* genome database
SH2	Src-homology-2
SILAC	Stable isotope labeling by amino acids in cell culture
SPAD	Signaling pathway database
Ssb	Single-strand binding proteins
STAT	Signal transducers and activators of transcription
STKE	Signal transduction knowledge environment
SuccCoA	Succinyl-coenzyme A
SUCD4	Succinate dehydrogenase
SVD	Singular value decomposition
T3P	Triose 3-phosphate
TAIS	Target-assisted iterative screening
TALA	Transaldolase
TCA	Tricarboxylic acid
TF	Transcription factor
TGF	Transforming growth factor
THD2	NAD(P) transhydrogenase
TIGR	Institute for genomic research
TKT	Transketolase
TPI	Triose-phosphate isomerase
TRANSFAC	Transcription factor database
TrEMBL	Translated European Molecular Biology Laboratory
tRNA	Transfer ribonucleic acid
UNG	Uracil glycosylase
VSR	DNA endonucleus
YPD	Yeast protein database

Escherichia coli Core Metabolic Network

Table B.1: A list of compounds in the core *Escherichia coli* network, their abbreviations, and cellular location.

Abbreviation	Name	Compartment
13dpg	3-phospho-D-glyceroyl phosphate	Cytosol
2pg	D-glycerate 2-phosphate	Cytosol
3pg	3-phospho-D-glycerate	Cytosol
6pgc	6-phospho-D-gluconate	Cytosol
6pgl	6-phospho-D-glucono-1,5-lactone	Cytosol
ac	Acetate	Cytosol
ac[e]	Acetate	Extracellular
accoa	Acetyl-CoA	Cytosol
actp	Acetyl phosphate	Cytosol
adp	ADP	Cytosol
akg	2-oxoglutarate	Cytosol
akg[e]	2-oxoglutarate	Extracellular
amp	AMP	Cytosol
atp	ATP	Cytosol
cit	Citrate	Cytosol
co2	CO2	Cytosol
co2[e]	CO2	Extracellular
coa	Coenzyme A	Cytosol
dhap	Dihydroxyacetone phosphate	Cytosol
e4p	D-erythrose 4-phosphate	Cytosol
etoh	Ethanol	Cytosol
etoh[e]	Ethanol	Extracellular
f6p	D-fructose 6-phosphate	Cytosol
fad	FAD	Cytosol
fadh2	FADH2	Cytosol
fdp	D-fructose 1,6-bisphosphate	Cytosol
for	Formate	Cytosol

(continued)

Table B.1 (continued).

Abbreviation	Name	Compartment
for[e]	Formate	Extracellular
fum	Fumarate	Cytosol
fum[e]	Fumarate	Extracellular
g3p	Glyceraldehyde 3-phosphate	Cytosol
g6p	D-glucose 6-phosphate	Cytosol
glc-D[e]	D-glucose	Extracellular
glx	Glyoxylate	Cytosol
h	H^+	Cytosol
h2o	H_2O	Cytosol
h2o[e]	H_2O	Extracellular
h[e]	H^+	Extracellular
icit	Isocitrate	Cytosol
lac-D	D-lactate	Cytosol
lac-D[e]	D-lactate	Extracellular
mal-L	L-malate	Cytosol
nad	Nicotinamide adenine dinucleotide	Cytosol
nadh	Nicotinamide adenine dinucleotide − reduced	Cytosol
nadp	Nicotinamide adenine dinucleotide phosphate	Cytosol
nadph	Nicotinamide adenine dinucleotide phosphate − reduced	Cytosol
o2	O_2	Cytosol
o2[e]	O_2	Extracellular
oaa	Oxaloacetate	Cytosol
pep	Phosphoenolpyruvate	Cytosol
pi	Phosphate	Cytosol
pi[e]	Phosphate	Extracellular
pyr	Pyruvate	Cytosol
pyr[e]	Pyruvate	Extracellular
q8	Ubiquinone-8	Cytosol
q8h2	Ubiquinol-8	Cytosol
r5p	Alpha-D-ribose 5-phosphate	Cytosol
ru5p-D	D-ribulose 5-phosphate	Cytosol
s7p	Sedoheptulose 7-phosphate	Cytosol
succ	Succinate	Cytosol
succ[e]	Succinate	Extracellular
succoa	Succinyl-CoA	Cytosol
xu5p-D	D-xylulose 5-phosphate	Cytosol

Table B.2: A list of reactions in the core *E. coli* network and associated chemical, genetic, and classification information.

Abbreviation	Official name	Equation	Subsystem	Gene description	Protein description	Protein class description
ENO	Enolase	[c] : 2pg ↔ h2o + pep	Glycolysis/gluconeogenesis	b2779	Eno	EC-4.2.1.11
FBA	Fructose-bisphosphate aldolase	[c] : fdp ↔ dhap + g3p	Glycolysis/gluconeogenesis	b2925	FbaA	EC-4.1.2.13
FBP	Fructose-bisphosphatase	[c] : fdp + h2o → f6p + pi	Glycolysis/gluconeogenesis	b4232	Fbp	EC-3.1.3.11
GAPD	Glyceraldehyde-3-phosphate dehydrogenase	[c] : g3p + nad + pi ↔ 13dpg + h + nadh	Glycolysis/gluconeogenesis	b1779, b1416, b1417	GapA, GapC	EC-1.2.1.12
PDH	Pyruvate dehydrogenase	[c] : coa + nad + pyr → accoa + co2 + nadh	Glycolysis/gluconeogenesis	b0114, b0115, b0116	AceE	EC-undetermined
PFK	Phosphofructokinase	[c] : atp + f6p → adp + fdp + h	Glycolysis/gluconeogenesis	b3916, b1723	PfkA, PfkB	EC-2.7.1.11
PGI	Glucose-6-phosphate isomerase	[c] : g6p ↔ f6p	Glycolysis/gluconeogenesis	b4025	Pgi	EC-5.3.1.9
PGK	Phosphoglycerate kinase	[c] : 3pg + atp ↔ 13dpg + adp	Glycolysis/gluconeogenesis	b2926	Pgk	EC-2.7.2.3
PGM	Phosphoglycerate mutase	[c] : 2pg ↔ 3pg	Glycolysis/gluconeogenesis	b4395, b0755	GpmB, GpmA	EC-5.4.2.1
PPS	Phosphoenolpyruvate synthase	[c] : atp + h2o + pyr → amp + (2) h + pep + pi	Glycolysis/gluconeogenesis	b1702	Ppsa	EC-2.7.9.2
PYK	Pyruvate kinase	[c] : adp + h + pep → atp + pyr	Glycolysis/gluconeogenesis	b1854, b1676	Pyka, Pykf	EC-2.7.1.40
TPI	Triose-phosphate isomerase	[c] : dhap ↔ g3p	Glycolysis/gluconeogenesis	b3919	Tpi	EC-5.3.1.1
G6PDH2r	Glucose 6-phosphate dehydrogenase	[c] : g6p + nadp ↔ 6pgl + h + nadph	Pentose phosphate cycle	b1852	Zwf	EC-1.1.1.49
GND	Phosphogluconate dehydrogenase	[c] : 6pgc + nadp → co2 + nadph + ru5p-D	Pentose phosphate cycle	b2029	Gnd	EC-1.1.1.44
PGL	6-phosphogluconolactonase	[c] : 6pgl + h2o → 6pgc + h	Pentose phosphate cycle			EC-3.1.1.31
RPE	Ribulose 5-phosphate 3-epimerase	[c] : ru5p-D ↔ xu5p-D	Pentose phosphate cycle	b3386	Rpeec	EC-5.1.3.1

Abbrev.	Name	Reaction	Pathway	Gene loci	Gene	EC number
RPI	Ribose-5-phosphate isomerase	[c] : r5p ↔ ru5p-D	Pentose phosphate cycle	b2914, b4090	RpiA, RpiB	EC-5.3.1.6
TALA	Transaldolase	[c] : g3p + s7p ↔ e4p + f6p	Pentose phosphate cycle	b0008	TalB	EC-2.2.1.2
TKT1	Transketolase	[c] : r5p + xu5p-D ↔ g3p + s7p	Pentose phosphate cycle	b2935, b2465	TktA, TktB	EC-2.2.1.1
TKT2	Transketolase	[c] : e4p + xu5p-D ↔ f6p + g3p	Pentose phosphate cycle	b2935, b2465	TktA, TktB	EC-2.2.1.1
ADK1	Adenylate kinase	[c] : amp + atp ↔ (2) adp	Oxidative phosphorylation	b0474	Adk	EC-2.7.4.3
ATPM	ATP maintenance requirement	[c] : atp + h2o → adp + h + pi	Oxidative phosphorylation			
ATPS4r	ATP synthase (four protons for one ATP)	adp[c] + (4) h[e] + pi[c] ↔ atp[c] + (3) h[c] + h2o[c]	Oxidative phosphorylation	b3736, b3737, b3738, b3731, b3732, b3733	AtpF0, AtpF1, AtpI	EC-3.6.3.14
CYTBD	Cytochrome oxidase bd (ubiquinol-8: 2 protons)	(2) h[c] + (0.5) o2[c] + q8h2[c] → (2) h[e] + h2o[c] + q8[c]	Oxidative phosphorylation	b0429, b0430, b0431, b0432	CyoA	EC-1.10.2.2
NADH11	NADH dehydrogenase (ubiquinone-8 8 2 protons)	(3) h[c] + nadh[c] + q8[c] → (2) h[e] + nad[c] + q8h2[c]	Oxidative phosphorolation	b2276, b2277, b2278, b2279, b2280, b2281, b2282, b2283, b2284, b2285, b2286, b2287, b2288	Nuo	EC-1.6.5.3
SUCD4	Succinate dehydrogenase	[c] : fadh2 + q8 ↔ fad + q8h2	Oxidative phosphorylation	b0721, b0722, b0723, b0724	Sdh	EC-undetermined
NADTRHD	NAD transhydrogenase	[c] : nad + nadph → nadh + nadp	Oxidative phosphorylation	b1602, b1603	Pnt	EC-1.6.1.2
THD2	NAD(P) transhydrogenase	(2) h[e] + nadh[c] + nadp[c] → (2) h[c] + nad[c] + nadph[c]	Oxidative phosphorylation	b1602, b1603	Pnt	EC-1.6.1.1
ACKr	Acetate kinase	[c] : ac + atp ↔ actp + adp	Pyruvate metabolism	b2296	AckA	EC-2.7.2.1
ADHEr	Acetaldehyde dehydrogenase	[c] : accoa + (2) h + (2) nadh ↔ coa + etoh + (2) nad	Pyruvate metabolism	b1241	AdhE	EC-1.1.1.1, EC-1.2.1.10
LDH_D	D-lactate dehydrogenase	[c] : lac-D + nad ↔ h + nadh + pyr	Pyruvate metabolism	b1380	Ldh	EC-1.1.1.28

(continued)

Table B.3 (continued).

Abbreviation	Official name	Equation	Subsystem	Gene description	Protein description	Protein class description
PFL	Pyruvate formate lyase	[c] : coa + pyr → accoa + for	Pyruvate metabolism	b3951, b3952	PflC	EC-2.3.1.54
PTAr	Phosphotransacetylase	[c] : accoa + pi ↔ actp + coa	Pyruvate metabolism	b2297	Pta	EC-2.3.1.8
ICL	Isocitrate lyase	[c] : icit → glx + succ	Anaplerotic reactions	b4015	AceA	EC-4.1.3.1
MALS	Malate synthase	[c] : accoa + glx + h2o → coa + h + mal-L	Anaplerotic reactions	b4014, b2976	AceB, GlcB	EC-4.1.3.2
ME1	Malic enzyme (NAD)	[c] : mal-L + nad → co2 + nadh + pyr	Anaplerotic reactions	b1479	Sfc	EC-1.1.1.38
ME2	Malic enzyme (NADP)	[c] : mal-L + nadp → co2 + nadph + pyr	Anaplerotic reactions	b2463	Mae	EC-1.1.1.40
PPC	Phosphoenolpyruvate carboxylase	[c] : co2 + h2o + pep → h + oaa + pi	Anaplerotic reactions	b3956	Ppc	EC-4.1.1.31
PPCK	Phosphoenolpyruvate carboxykinase	[c] : atp + oaa → adp + co2 + pep	Anaplerotic reactions	b3403	Pck	EC-4.1.1.49
AKGt2r	2-oxoglutarate reversible transport via symport	akg[e] + h[e] ↔ akg[c] + h[c]	Transport	b2587	KgtPec	
CO2t	CO2 transporter via diffusion	co2[e] ↔ co2[c]	Transport			
D-LACt2	D-lactate transport via proton symport	h[e] + lac-D[e] ↔ h[c] + lac-D[c]	Transport			
ETOH2r	Ethanol reversible transport via proton symport	etoh[e] + h[e] ↔ etoh[c] + h[c]	Transport			
FORt	Formate transport via diffusion	for[e] ↔ for[c]	Transport			
GLCpts	D-glucose transport via PEP:Pyr PTS	glc-D[e] + pep[c] → g6p[c] + pyr[c]	Transport	b1101, b2415, b2416, b2417	Pts	
H2Ot	H2O transport via diffusion	h2o[e] ↔ h2o[c]	Transport			

Bibliography

[1] B. Alberts, D. Bray, J. Lewis, M. Raff, K. Roberts, and J.D. Watson. *Molecular Biology of the Cell*. Garland Publishing, New York, 2002.

[2] E. Almaas, B. Kovács, T. Vicsek, Z.N. Oltvai, and A.-L. Barabási. Global organization of metabolic fluxes in the bacterium *Escherichia coli*. *Nature*, 427:839–843, 2004.

[3] O. Alter, P.O. Brown, and D. Botstein. Singular value decomposition for genome-wide expression data processing and modeling. *Proceedings of the National Academy of Sciences of the United States of America*, 97:10101–10106, 2000.

[4] D.E. Atkinson. *Cellular energy metabolism and its regulation*. Academic Press, New York, 1977.

[5] J.E. Bailey. Mathematical modeling and analysis in biochemical engineering: Past accomplishments and future opportunities. *Biotechnology Progress*, 14:8–20, 1998.

[6] J.E. Bailey. Lessons from metabolic engineering for functional genomics and drug discovery. *Nature Biotechnology*, 17:616–618, 1999.

[7] A.L. Barabasi. *Linked: The New Science of Networks*. Perseus, Cambridge, 2002.

[8] M.S. Bazaraa, H.D. Sherali, and C.M. Shetty. *Nonlinear Programming: Theory and Algorithms*, 2nd edition. John Wiley and Sons, NJ, Hoboken, 1993.

[9] D.A. Beard and H. Qian. Thermodynamic-based computational profiling of cellular regulatory control in hepatocyte metabolism. *American Journal of Physiology*, 288:E633–E644, 2005.

[10] S.A. Becker and B.O. Palsson. Genome-scale reconstruction of the metabolic network in *Staphylococcus aureus* N315: An initial draft to the two-dimensional annotation. *BMC Microbiology*, 5:8–19, 2005.

[11] S.A. Becker, N.D. Price, and B.O. Palsson. Metabolite coupling in genome-scale metabolic networks. In preparation.

[12] J.R. Beckwith. Regulation of the *lac* operon. Recent studies on the regulation of lactose metabolism in *Escherichia coli* support the operon model. *Science*, 156:597–604, 1967.

[13] S.L. Bell and B.O. Palsson. ExPa: A program for calculating extreme pathways in biochemical reaction networks. *Bioinformatics*, 21:1739–1740, 2005.

[14] F. Berger, M.H. Ramirez-Harmdez, and M. Ziegler. The new life of a centenarian: Signalling functions of NAD(P). *Trends in Biochemical Sciences*, 29:111–118, 2004.

[15] M.A. Bhatti. *Practical Optimization Methods: With Mathematica Applications*. Springer, New York, 2000.

[16] B. Blagoev, S.-E. One, I. Kratchmarova, and M. Mann. Temporal analysis of phosphotyrosine-dependent signaling networks by quantitative proteomics. *Nature Biotechnology*, 22:1139–1145, 2004.

[17] H.P.J. Bonarius, G. Schmid, and J. Tramper. Flux analysis of underdetermined metabolic networks: The quest for the missing constraints. *Trends in Biotechnology*, 15:308–314, 1997.

[18] M. Boutros, A.A. Kiger, S. Aamknecht, et al. Genome-wide RNAi analysis of growth and viability in Drosophila cells. *Science*, 303:832–835, 2004.

[19] J.M. Bower and H. Bolouri (ed). *Computational Modeling of Genetic and Biochemical Networks*. MIT Press, Cambridge, MA, 2004.

[20] T.R. Brummelkamp, S.M.S Nijman, A.M.G. Dirac, and R. Bernaerg. Loss of the cylindromatosis tumour suppressor inhibits apoptosis by activating NF-$\kappa\beta$. *Nature*, 424:797–801, 2003.

[21] M.L. Bulyk, X. Huang, Y. Choo, and G.M. Church. Exploring the DNA-binding specificities of zinc fingers with DNA microarrays. *Proceedings of the National Academy of Sciences of the United States of America*, 98:7158–7163, 2001.

[22] M. Bunemann, M. Frank, and M.J. Lohse. Gi protein activation in intact cells involves subunit rearrangement rather than dissociation. *Proceedings of the National Academy of Sciences of the United States of America*, 100:16077–16082, 2003.

[23] A.P. Burgard and C.D. Maranas. An optimization based framework for inferring and testing hypothesized metabolic objective functions. *Biotechnology and Bioengineering*, 82:670–677, 2003.

[24] A.P. Burgard, E.V. Nikolaev, C.H. Schilling, and C.D. Maranas. Flux coupling analysis of genome-scale metabolic network reconstructions. *Genome Research*, 14:301–312, 2004.

[25] A.P. Burgard, P. Pharkya, and C.D. Maranas. Optknock: A bilevel programming framework for identifying gene knockout strategies for microbial strain optimization. *Biotechnology and Bioengineering*, 84:647–657, 2003.

[26] W. Busch and M.H. Saier. The IUBMB-endorsed transporter classification system. *Molecular Biotechnology*, 27(3):253–262, 2004.

[27] L.C. Cantley. The phosphoinositide 3-kinase pathway. *Science*, 296:1655–1657, 2002.

[28] R. Carlson, D. Fell, and F. Srienc. Metabolic pathway analysis of a recombinant yeast for rational strain development. *Biotechnology and Bioengineering*, 79:121–134, 2002.

[29] I. Cases, V. de Lorenzo, and C.A. Ouzounis. Transcription regulation and environmental adaptation in bacteria. *Trends in Microbiology*, 11:248–253, 2003.

[30] V. Chvatal. *Linear Programming*. W.H. Freeman, New York, 1983.

[31] B.L. Clarke. Stability of complex reaction networks. *Advances in Chemical Physics*. John Wiley, and Sons, New York, 1980.

[32] International Human Genome Sequencing Consortium. Finishing the euchromatic sequence of the human genome. *Nature*, 431:931–945, 2004.

[33] M. Covert, M.J. Herrgard, J.L. Reed, and B.O. Palsson. Integrating high-throughput data and computational models leads to *E.coli* network elucidation. *Nature*, 429:92–96, 2004.

[34] M.W. Covert, I. Famili, and B.O. Palsson. Identifying constraints that govern cell behavior: A key to converting conceptual to computational models in biology? *Biotechnology and Bioengineering*, 84:763–772, 2003.

[35] M.W. Covert and B.O. Palsson. Transcriptional regulation in constraints-based metabolic models of *Escherichia coli. Journal of Biological Chemistry*, 277:28058–28064, 2002.

[36] M.W. Covert and B.O. Palsson. Constraints-based models: Regulation of gene expression reduces the steady-state solution space. *Journal of Theoretical Biology*, 221:309–325, 2003.

[37] M.W. Covert, C.H. Schilling, I. Famili, et al. Metabolic modeling of microbial stains *in silico. Trends in Biochemical Sciences*, 26:179–186, 2001.

[38] M.W. Covert, C.H. Schilling, and B.O. Palsson. Regulation of gene expression in flux balance models of metabolism. *Journal of Theoretical Biology*, 213:73–88, 2001.

[39] F. Crick. *What Mad Pursuit: A Personal View of Scientific Discovery*. Basic Books, New York, 1988.

[40] C. Csank, M.C. Costanzo, J. Hirschman, et al. Three yeast proteome databases: YPD, PombePD, and CalPD (MycoPathPD). *Methods in Enzymology*, 350:347–373, 2002.

[41] A. Danchin. *The Delphic Boat: What Genomes Tell Us*. Harvard University Press, Cambridge, MA, 2003.

[42] T. Dandekar, S. Schuster, B. Snel, M. Huynen, and P. Bork. Pathway alignment: Application to the comparative analysis of glycolytic enzymes. *Biochemical Journal*, 343:115–124, 1999.

[43] H. David, M. Akesson, and J. Nielsen. Reconstruction of the central carbon metabolism of *Aspergillus niger. European Journal of Biochemistry*, 270:4243–4253, 2003.

[44] F. Di Lisa, C.Z. Fan, G. Gambassi, B.A. Hogue, I. Kudryashova, and R.G. Hansford. Altered pyruvate dehydrogenase control and mitochondrial free Ca^{2+} in hearts of cardiomyopathic hamsters. *American Journal of Physiology*, 264:2188–2197, 1993.

[45] H. Dihazi, R. Kessler, and K. Eschrich. High osmolarity glycerol (hog) pathway-induced phosphorylation and activation of 6-phosphofructo-2-kinase are essential for glycerol accumulation and yeast cell proliferation

under hyperosmotic stress. *Journal of Biological Chemistry*, 279:23961–23968, 2004.

[46] D.Q. Ding, y. Tomita, A. Yamamoto, Y. Chikashiue, T. Haracuchi, Y. Hiraoka. Large-scale screening of intracellular protein localization in living fission yeast cells by the use of a GFP-fusion genomic DNA library. *Genes Cells*, 5:169–190, 2000.

[47] N.C. Duarte, M.J. Herrgard, and B.O. Palsson. Reconstruction and validation of *Saccharomyces cerevisiae* iND750, a fully compartmentalized genome-scale metabolic model. *Genome Research*, 14:1298–1309, 2004.

[48] J.S. Edwards, R. Ramakrishna, and B.O. Palsson. Characterizing the metabolic phenotype: A phenotype phase plane analysis. *Biotechnology and Bioengineering*, 77:27–36, 2002.

[49] J.S. Edwards, R.U. Ibarra, and B.O. Palsson. *In silico* predictions of *Escherichia coli* metabolic capabilities are consistent with experimental data. *Nature Biotechnology*, 19:125–130, 2001.

[50] J.S. Edwards and B.O. Palsson. Systems properties of the *Haemophilus influenzae* Rd metabolic genotype. *Journal of Biological Chemistry*, 274:17410–17416, 1999.

[51] J.S. Edwards and B.O. Palsson. The *Escherichia coli* MG1655 *in silico* metabolic genotype; its definition, characteristics, and capabilities. *Proceedings of the National Academy of Sciences of the United States of America*, 97:5528–5533, 2000.

[52] C.P. Fall, E.S. Marland, J.M. Wagner, and J.J. Tyson (eds). *Computational Cell Biology Series: Interdisciplinary Applied Mathematics*, vol. 20. Springer-Verlag Telos, New York, 2002.

[53] I. Famili. *Systemic Analysis of the Stoichiometric Matrix and Kinetic Characterization of Metabolic Networks*. PhD thesis, University of California, San Diego, 2004.

[54] I. Famili, J. Forster, J. Nielsen, and B.O. Palsson. *Saccharomyces cerevisiae* phenotypes can be predicted using constraint-based analysis of a genome-scale reconstructed metabolic network. *Proceedings of the National Academy of Sciences of the United States of America*, 100:13134–13139, 2003.

[55] I. Famili, R. Mahadevan, and B.O. Palsson. K-cone analysis: Determining all candidate values for kinetic parameters on a network-scale. *Biophysical Journal*, 88:1616–1625, 2005.

[56] I. Famili and B.O. Palsson. The convex basis of the left null space of the stoichiometric matrix leads to the definition of metabolically meaningful pools. *Biophysical Journal*, 85:16–26, 2003.

[57] I. Famili and B.O. Palsson. Systemic metabolic reactions are obtained by singular value decomposition of genome-scale stoichiometric matrices. *Journal of Theoretical Biology*, 224:87–96, 2003.

[58] R.P. Faynman. *The Character of Physical Law*. MIT Press, Cambridge, MA, 1965.

[59] D.A. Fell. Metabolic control analysis: A survey of its theoretical and experimental development. *Biochemical Journal*, 286:313–330, 1992.

[60] R.D. Fleischmann, M.D. Adamas, D. White, et al. Whole-genome random sequencing and assembly of *Haemophilus influenzae* Rd. *Science*, 269:496–498, 1995.

[61] S.S. Fong. *Adaptive Evolution of Escherichia coli: Systems Biology and a Genome-Scale Metabolic Model*. PhD thesis, University of California, San Diego, 2004.

[62] S.S. Fong, A.P. Burgard, C.D. Herring, et al. *In silico* design and adaptive evolution of *Escherichia coli* for production of lactic acid. *Biotechnology and Bioengineering*, 91: 643–648 (2005).

[63] S.S. Fong, J.Y. Marciniak, and B.O. Palsson. Description and interpretation of adaptive evolution of *Escherichia coli* K12 MG1655 using a genome-scale *in silico* metabolic model. *Journal of Bacteriology*, 185:6400–6408, 2003.

[64] S.S. Fong and B.O. Palsson. Metabolic gene-deletion strains of *Escherichia coli* evolve to computationally predicted growth phenotypes. *Nature Genetics*, 36:1056–1058, 2004.

[65] A.R. Forrest and T. Ravasi. Phosphoregulators: Protein kinases and protein phosphatases of mouse. *Genome Research*, 13:1443–1454, 2003.

[66] J. Forster, I. Famili, P.C. Fu, B.O. Palsson, and J. Nielsen. Genome-scale reconstruction of the *Saccharomyces cerevisiae* metabolic network. *Genome Research*, 13:244–253, 2003.

[67] J. Forster, I. Famili, B.O. Palsson, and J. Nielsen. Large-scale evaluation of *in silico* gene deletions in *Saccharomyces cerevisiae*. *OMICS*, 7(2):193–202, 2003.

[68] M. Fussenegger, J.E. Bailey, and J. Varner. A mathematical model of caspase function in apoptosis. *Nature Biotechnology*, 18:768–774, 2000.

[69] D. Garfinkel, L. Garfinkel, M. Pring, S.B. Green, and B. Chance. *Annual Review of Biochemistry*, 39:473–498, 1970.

[70] A.C. Gavin, M. Bosche, R. Krause, et al. Functional organization of the yeast proteome by systematic analysis of protein complexes. *Nature*, 415:141–147, 2002.

[71] S.Y. Gerdes, M.D. Scholle, J.W. Campbell, et al. Experimental determination and system level analysis of essential genes in *Escherichia coli* MG1655. *Journal of Bacteriology*, 185:5673–5684, 2003.

[72] J. Gerhart. 1998 Warkany lecture: Signaling pathways in development. *Teratology*, 60:226–239, 1999.

[73] J. Gerhart and M. Kirschner (eds). *Cells, Embryos, and Evolution: Toward a Cellular and Developmental Understanding of Phenotypic Variation and Evolutionary Adaptability*. Blackwell Science, Malden, MA, 1997.

[74] A. Gilman, M.I. Simon, H.R. Bourne, et al. Overview of the alliance for cellular signaling. *Nature*, 420:703–706, 2002.

[75] A. Goldbeter and P. Nicolis. An allosteric enzyme model with positive feedback applied to glycolytic oscillations. *Progress in Theoretical Biology*, 4:65–160, 1976.

[76] D.S. Goodsell. *The Machinery of Life*. Springer-Verlag, New York, 1993.

[77] D.S. Goodsell. *Our Molecular Nature: The Body's Motors, Machines and Messages*. Copernicus, New York, 1997.

[78] B.C. Goodwin. *Oscillatory organization in cells, a dynamic theory of cellular control processes*. Academic Press, New York, 1963.

[79] A. Grigoriev. On the number of protein–protein interactions in the yeast proteome. *Nucleic Acids Research*, 31:4157–4161, 2003.

[80] N. Guelzim, S. Bottani, P. Bourgine, and F. Kepes. Topological and causal structure of the yeast transcriptional regulatory network. *Nature Genetics*, 31:60–63, 2002.

[81] D.A. Hall, H. Zhu, X. Zhu, T. Royce, M. Gerstein, and M. Snyder. Regulation of gene expression by a metabolic enzyme. *Science*, 306:482–484, 2004.

[82] C.T. Harbison, D.B. Gordon, T.I. Lee, et al. Transcriptional regulatory code of a eukaryotic genome. *Nature*, 431:99–104, 2004.

[83] G.W. Hatfield and C.J. Benham. DNA topology-mediated control of global gene expression in *Escherichia coli*. *Annual Review of Genetics*, 36:175–203, 2002.

[84] R.H. Herman. The principles of metabolic control. In *The Principles of Metabolic Control in Mammalian Systems* (R.H. Herman, R.M. Cohn, and P.D. Mcnamara, eds.). Plenum Press, New York, 1980.

[85] M.J. Herrgard, M.W. Covert, and B.O. Palsson. Reconstruction of microbial transcriptional regulatory networks. *Current Opinion in Biotechnology*, 15:70–77, 2004.

[86] M.J. Herrgard. *Reconstruction and Systems Analysis of Genome-Scale Metabolic and Regulatory Networks in Saccharomyces cerevisiae*. PhD thesis, University of California, San Diego, 2004.

[87] M.J. Herrgard, M.W. Covert, and B.O. Palsson. Reconciling gene expression data with known genome-scale regulatory network structure. *Genome Research*, 13:2423–2434, 2003.

[88] M.J. Herrgard, B.S. Lee, and B.O. Palsson. Integrated analysis of regulatory and metabolic networks reveals novel regulatory mechanisms in *Saccharomyces cerevisiae*. *Genome Research*, Submitted:x–x, 2005. Still Submitted

[89] Y. Ho, A. Gruhler, A. Heilbut, et al. Systematic identification of protein complexes in *Saccharomyces cerevisiae* by mass spectrometry. *Nature*, 415: 180–183, 2002.

[90] P.S. Hoffman, A. Goodwin, J. Johnsen, K. Magee, and S.J. Veldhuyzen van Zanten. Metabolic activities of metronidazole-sensitive and -resistant strains of *Helicobacter pylori*: Repression of pyruvate oxidoreductase and expression of isocitrate lyase activity correlate with resistance. *Journal of Bacteriology*, 178:4822–4829, 1996.

[91] N.S. Holter, M. Mitra, A. Maritan, M. Cieplak, J.R. Banavar, and N.V. Fedoroff. Fundamental patterns underlying gene expression profiles: Simplicity from complexity. *Proceedings of the National Academy of Sciences of the United States of America*, 97:8409–8414, 2000.

[92] S. Holzhutter and H.G. Holzhutter. Computational design of reduced metabolic networks. *ChemBioChem*, 5:1401–1422, 2004.

[93] S.H. Hong, J.S. Kim, S.Y. Lee, et al. The genome sequence of the capnophilic rumen bacterium *Mannheimia succiniciproducens*. *Nature Biotechnology*, 22:1275–1281, 2004.

[94] L.Hood and D. Galas. The digital code of DNA. *Nature*, 421:444–448, 2003.

[95] L. Hood and R.M. Perlmutter. The impact of systems approaches on biological problems in drug discovery. *Nature Biotechnology*, 22:1215–1217, 2004.

[96] M.L. Howard and E.H. Davidson. *cis*-regulatory control circuits in development. *Developmental Biology*, 271(1):109–118, 2004.

[97] T.R. Hughes, M.J. Martin, A.R. Jones, et al. Functional discovery via a compendium of expression profiles. *Cell*, 102:109–126, 2000.

[98] W.K. Huh, J.V. Falvo, L.C. Gerke, et al. Global analysis of protein localization in budding yeast. *Nature*, 425:686–691, 2003.

[99] R.U. Ibarra, J.S. Edwards, and B.O. Palsson. *Escherichia coli* K-12 undergoes adaptive evolution to achieve *in silico* predicted optimal growth. *Nature*, 420:186–189, 2002.

[100] T. Ideker, V.T. Horsson, J.A. Ranish, et al. Integrated genomic and proteomic analyses of a systematically perturbed metabolic network. *Science*, 292:929–934, 2001.

[101] T. Ideker, T. Galitski, and L. Hood. A new approach to decoding life: Systems biology. *Annual Review of Genomics and Human Genetics*, 2:343–372, 2001.

[102] F. Jacob. Evolution and tinkering. *Science*, 196:1161–1166, 1977.

[103] N. Jamshidi, S.J. Wiback, and B.O. Palsson. *In silico* model-driven assessment of the effects of single nucleotide polymorphisms (SNPs) on human red blood cell metabolism. *Genome Research*, 12:1687–1692, 2002.

[104] H. Jeong, S.P. Mason, A.L. Barabasi, and Z.N. Oltvai. Lethality and centrality in protein networks. *Nature*, 411:41–42, 2001.

[105] H. Jeong, B. Tombor, R. Albert, Z. N. Oltvai, and A.L. Barabasi. The large-scale organization of metabolic networks. *Nature*, 407:651–654, 2000.

[106] A. Joshi and B.O. Palsson. Metabolic dynamics in the human red cell, Part I – a comprehensive model, Part II – interactions with the environment. *Journal of Theoretical Biology*, 141:515–545, 1989.

[107] M. Kaern, et al. Stochasticity in gene expression: From theories to phenotypes. *Nature Reviews Genetics*, May 10:1–14, 2005.

[108] P. Karp, M. Krummenacker, S. Paley, and J. Wagg. Integrated pathway-genome databases and their role in drug discovery. *Trends in Biotechnology*, 17:275–281, 1999.

[109] K.J. Kauffman, J.D. Pajerowski, N. Jamshidi, B.O. Palsson, and J.S. Edwards. Description and analysis of metabolic connectivity and dynamics in the human red blood cell. *Biophysical Journal*, 83:646–662, 2002.

[110] K.J. Kauffman, P. Prakash, and J.S. Edwards. Advances in flux balance analysis. *Current Opinion in Biotechnology*, 14:491–496, 2003.

[111] D.E. Kaufman and R.L. Smith. Direction choice for accelerated convergence in hit-and-run sampling. *Operations Research*, 46:84–95, 1998.

[112] J.K. Kelleher. Flux estimation using isotopic tracers: Common ground for metabolic physiology and metabolic engineering. *Metabolic Engineering*, 3:100–110, 2001.

[113] J. Kim and C.V. Dang. Multifaceted roles of glycolytic enzymes. *Trends in Biochemical Sciences*, 30:142–150, 2005.

[114] F.C. Knowles, J.D. Chanley, and N.G. Pon. Spectral changes arising from the action of spinach chloroplast ribosephosphate isomerase on ribose 5–phosphate. *Archives of Biochemist and Biophysics.*, 202:106–115, 1980.

[115] S.D. Kobayashi, S.D. Ehrlkh, A. Albemini, et al. Essential *Bacillus subtilis* genes. *Proceedings of the National Academy of Sciences of the United States of America*, 100:4678–4683, 2003.

[116] A. Kumar and A. Snyder. Emerging technologies in yeast genomics. *Nature Reviews Genetics*, 2:302–312, 2001.

[117] E.S. Lander, L.M. Linton, B. Birren, et al. Initial sequencing and analysis of the human genome. *Nature*, 409:860–921, 2001.

[118] M.T. Laub, H.H. McAdams, T. Feldblyum, C.M. Fraser, and L. Shapiro. Global analysis of the genetic network controlling a bacterial cell cycle. *Science*, 290:2144–2148, 2000.

[119] S. Lee, C. Phalakornkule, M.M. Domach, and I.E. Grossmann. Recursive MILP model for finding all the alternate optima in LP models for metabolic networks. *Computers and Chemical Engineering*, 24:711–716, 2000.

[120] A.L. Lehninger, D.L. Nelson, and M.M. Cox. *Principles of Biochemistry*. Worth Publishers, New York, 1993.

[121] J.C. Liao, S.Y. Hou, and Y.P. Chao. Pathway analysis, engineering and physiological considerations for redirecting central metabolism. *Biotechnology and Bioengineering*, 52:129–140, 1996.

[122] W.R. Loewenstein. *The Touchstone of Life*. Oxford University Press, Oxford, 1999.

[123] W. Loomis and S. Thomas. Kinetic analysis of biochemical differentiation in *Dictyostelium discoideum*. *Journal of Biological Chemistry*, 251:6252–6258, 1976.

[124] W.F. Loomis and B. Magasanik. Glucose–lactose diauxie in *Escherichia coli*. *Journal of Bacteriology*, 93:1397–1401, 1967.

[125] L. Lovasz. Hit-and-run mixes fast. *Mathematical Programming*, 86:443–461, 1999.

[126] L. Lum, S. Yao, B. Moler, et al. Identification of hedgehog pathway components by RNAi in Drosophila cultured cells. *Science*, 299:2039–2045, 2003.

[127] R. Mahadevan, et al. Model driven discovery of novel biological mechanisms in the metabolism of *Geobacter sulfurreducens*. In preparation.

[128] R. Mahadevan and B.O. Palsson. Properties of metabolic networks: Structure versus function. *Biophysical Journal*, 88:L7–L9, 2005.

[129] R.A. Majewski and M.M. Domach. Simple constrained optimization view of acetate overflow in *E. coli*. *Biotechnology and Bioengineering*, 35:732–738, 1990.

[130] G. Manning, D.B. Whyte, R. Martinez, T. Hunter and S. Sudarsanam. The protein kinase complement of the human genome. *Science*, 298:1912–1934, 2002.

[131] R. Marre (eds). *Legionella*. ASM Press, Washington, DC, 2001.

[132] M. Martin-Fernandez, O.T. Clarke, M.J. Tobin, S.U. Jones and G.R. Jones. Preformed oligomeric epidermal growth factor receptors undergo an ectodomain structure change during signaling. *Biophysical Journal*, 82:2415–2427, 2002.

[133] G. Matarese and A. La Cava. The intricate interface between immune system and metabolism. *Trends in Immunology*, 25:193–200, 2004.

[134] V. Matys, E. Fricke, R. Geffers, et al. TRANSFAC: Transcriptional regulation, from patterns to profiles. *Nucleic Acids Research*, 31:374–378, 2003.

[135] M.L. Mavrovouniotis, G. Stephanopoulos, and G. Stephanopoulos. Computer-aided synthesis of biochemical pathways. *Biotechnology and Bioengineering*, 36:1119–1132, 1990.

[136] E. Mayr. *This is Biology: the Science of the Living World*. Belknap Press of Harvard University Press, Cambridge, MA, 1997.

[137] T. Meyer and M.N. Teruel. Fluorescence imaging of signaling networks. *Trends in Cell Biology*, 13:101–106, 2003.

[138] H.L.T. Mobley, G.L. Menoz, and S.L. Hazell et al. (eds). *Helicobacter pylori: Physiology and Genetics*. ASM Press, Washington, DC, 2001.

[139] J. Monod, J. Wyman, and J.P. Changeaux. On the nature of allosteric transitions: A plausible model. *Journal of Molecular Biology*, 12:88–118, 1965.

[140] M.P. Murrell, K.J. Yarema, and A. Levchenko. The systems biology of glycosylation. *ChemBioChem*, 5:1334–1347, 2004.

[141] Buchler NE, U. Gerland, and T. Hwa. On schemes of combinatorial transcription logic. *Proceedings of the National Academy of Sciences of the United States of America*, 100:5136–5141, 2003.

[142] F.C. Neidhardt (ed). *Escherichia coli and Salmonella: Cellular and molecular biology*. ASM Press, Washington, DC, 1996.

[143] F.C. Neidhardt, J.L. Ingraham, and M. Schaechterm. *Physiology of the Bacterial Cell: A Molecular Approach*. Sinauer Associates, Sunderland, MA, 1990.

[144] J. Nielsen and J. Villadsen. *Bioreaction Engineering Principles*. Plenum Press, New York, 1994.

[145] U.B. Nielsen, M.H. Cardone, A.J. Sinskey, G. Macbeath, and P.K. Sorler et al. Profiling receptor tyrosine kinase activation by using ab microarrays. *Proceedings of the National Academy of Sciences, of the United States of America*, 100:9330–9335, 2003.

[146] B. Novak and J.J. Tyson. Quantitative analysis of a molecular model of mitotic control in fission yeast. *Journal Theoretical Biology*, 173:283–305, 1995.

[147] P. Oliveri and E.H. Davidson. Gene regulatory network controlling embryonic specification in the sea urchin. *Current Opinion in Genetics and Development*, 14(4):351–360, 2004.

[148] S.M. Paley and P.D. Karp. Evaluation of computational metabolic- pathway predictions for *Helicobacter pylori*. *Bioinformatics*, 18:715–724, 2002.

[149] B.O. Palsson and A. Joshi. On the dynamic orders of structured *E. coli* growth models. *Biotechnology and Bioengineering*.

[150] B.O. Palsson, A. Joshi, and S. S. Ozturk. Reducing complexity in metabolic networks: Making metabolic meshes manageable. *Federation Proceedings*, 46:2485, 1987.

[151] B.O. Palsson, J.C. Liao, and E.N. Lightfoot. Interpretation of biochemical dynamics using modal analysis. *Presented at the Annual AIChE Meeting, San Francisco, CA*, 1984.

[152] B.O. Palsson. The challenges of *in silico* biology. *Nature Biotechnology*, 18:1147–1150, 2000.

[153] B.O. Palsson. *In silico* biology through 'omics'. *Nature Biotechnology*, 20:649–650, 2002.

[154] B.O. Palsson. Methods of identifying therapeutic compounds in a genetically defined setting. Technical Report 6524797, United States Patent, 2003.

[155] B.O. Palsson. Two-dimensional annotation of genomes. *Nature Biotechnology*, 22:1218–1219, 2004.

[156] B.O. Palsson and S.N. Bhatia. *Tissue Engineering*. Prentice-Hall, Upper Sadole River, NJ, 2003.

[157] J.A. Papin. *Systems Analysis of Cellular Signaling Networks*. PhD thesis, University of California, San Diego, 2004.

[158] J.A. Papin, J.S. Telling, N.D. Pilke, S. Klamt, S. Schuster and B.O. Palsson et al. Comparison of network-based pathway analysis methods. *Trends in Biotechnology*, 22:400–405, 2004.

[159] J.A. Papin and B.O. Palsson. Hierarchical thinking in network biology: The unbiased modularization of biochemical networks. *Trends in Biochemical Sciences*, 29:641–647, 2004.

[160] J.A. Papin and B.O. Palsson. The JAK-STAT signaling network in the human β-cell: An extreme signaling pathway analysis. *Biophysical Journal*, 87:37–46, 2004.

[161] J.A. Papin and B.O. Palsson. Topological analysis of mass-balanced signaling networks: A framework to obtain emergent properties including crosstalk. *Journal of Theoretical Biology*, 227:283–297, 2004.

[162] J.A. Papin, N.D. Price, and B.O. Palsson. Extreme pathway lengths and reaction participation in genome-scale metabolic networks. *Genome Research*, 12:1889–1900, 2002.

[163] J.A. Papin, N.D. Price, S.J. Wiback, D.A. Fell, and B.O. Palsson. Metabolic pathways in the post-genome era. *Trends in Biochemical Sciences*, 28:250–258, 2003.

[164] C.S. Park and I.C. Schneider. Kinetic analysis of platelet-derived growth factor receptor/phosphoinositide 3-kinase/Akt signaling in fibroblasts. *Journal of Biological Chemistry*, 278:37064–37072, 2003.

[165] E. Perez-Rueda and J. Collado-Vides. The repertoire of DNA-binding transcriptional regulators in *Escherichia coli* K-12. *Nucleic Acids Research*, 28:1838–1847, 2000.

[166] T. Pfeiffer, I. Sanchez-Valdenebro, J.C. Nuno, F. Montero, and S. Schuster. Metatool: For studying metabolic networks. *Bioinformatics*, 15:251–257, 1999.

[167] R.D. Phair and T. Misteli. Kinetic modelling approaches to *in vivo* imaging. *Nature Reviews Molecular and Cellular Biology*, 2:898–907, 2001.

[168] C. Phalakornkule, S. Lee, T. Zhu, R. Koepsel, M.M. Ataai, I.E. Grossmann, and M.M. Domach. A MILP-based flux alternative generation and NMR experimental design strategy for metabolic engineering. *Metabolic Engineering*, 3:124–137, 2001.

[169] P. Pharkya, A.P. Burgard, and C.D. Maranas. Exploring the overproduction of amino acids using the bilevel optimization framework optKnock. *Biotechnology and Bioengineering*, 84:887–899, 2003.

[170] P. Pharkya, A.P. Burgard, and C.D. Maranas. OptStrain: A computational framework for redesign of microbial production systems. *Genome Research*, 14:2367–2376, 2004.

[171] J. Pramanik and J.D. Keasling. Stoichiometric model of *Escherichia coli* metabolism: Incorporation of growth-rate dependent biomass composition and mechanistic energy requirements. *Biotechnology and Bioengineering*, 56:398–421, 1997.

[172] J. Pramanik and J.D. Keasling. Effect of *Escherichia coli* biomass composition on central metabolic fluxes predicted by a stoichiometric model. *Biotechnology and Bioengineering*, 60:230–238, 1998.

[173] W.H. Press. *Numerical Recipes in C*. Cambridge University Press, New York, 1994.

[174] N.D. Price, I.F. Famili, D.A. Beard, and B.O. Palsson. Extreme pathways and Kirchhoff's second law. *Biophysical Journal*, 83:2879–2882, 2002.

[175] N.D. Price, J.A. Papin, C.H. Schilling, and B.O. Palsson. Genome-scale microbial *in silico* models: The constraints-based approach. *Trends in Biotechnology*, 21:162–169, 2003.

[176] N.D. Price, J.L. Reed, and B.O. Palsson. Genome-scale models of microbial cells: Evaluating the consequences of constraints. *Nature Reviews Microbiology*, 2:886–897, 2004.

[177] N.D. Price, J.L. Reed, J.A. Papin, I. Famili, and B.O. Palsson. Analysis of metabolic capabilities using singular value decomposition of extreme pathway matrices. *Biophysical Journal*, 84:794–804, 2003.

[178] N.D. Price, J.L. Reed, J.A. Papin, S.J. Wiback, and B.O. Palsson. Network-based analysis of regulation in the human red blood cell. *Journal of Theoretical Biology*, 225:1985–1994, 2003.

[179] N.D. Price, J. Schellenberger, and B.O. Palsson. Uniform sampling of steady state flux spaces: Means to design experiments and to interpret enzymopathies. *Biophysical Journal*, 87:2172–2186, 2004.

[180] M. Ptashne and A. Gann. *Genes and Signals*. Cold Spring Harbor Press, New York, 2002.

[181] R. Ramakrishna, J.S. Ednards, A. McCulloch, and B.O. Palsson. Flux balance analysis of mitochondrial energy metabolism: Consequences of systemic stoichiometric constraints. *The American Journal of Physiology: Regulatory, Integrative and Comparative Physiology*, 280:695–704, 2001.

[182] C.V. Rao, D.M. Wolf, and A.P. Arkin. Control, exploitation, and tolerance of intracellular noise. *Nature*, 420:231–237, 2002.

[183] J.L. Reed, and B.O. Palsson. Thirteen years of building constraints-based *in silico* models of *Escherichia coli*. *Journal of Bacteriology*, 185:2692–2699, 2003.

[184] J.L. Reed and B.O. Palsson. Genome-scale *in silico* models of *E. coli* have multiple equivalent phenotypic states: Assessment of correlated reaction subsets that comprise network states. *Genome Research*, 14:1797–1805, 2004.

[185] J.L. Reed, T.D. Vo, C.H. Schilling, and B.O. Palsson. An expanded genome-scale model of *Escherichia coli* K-12 (iJR904 GSM/GPR). *Genome Biology*, 4:R54.1–R54.12, 2003.

[186] J.G. Reich and E.E. Sel'kov (eds). *Energy Metabolism of the Cell*. Academic Press, New York, 1981.

[187] B. Ren, F. Robert, and J.J. Wyrick. Genome-wide location and function of DNA binding proteins. *Science*, 290:2306–2309, 2000.

[188] N. Rosenfeld, M.B. Elowitz, and U. Alon. Negative autoregulation speeds the response times of transcription networks. *Journal of Molecular Biology*, 323:785–793, 2002.

[189] M. Sato, T. Ozawa, K. Inukai, T. Asano, and Y. Umezawa. Fluorescent indicators for imaging protein phosphorylation in single living cells. *Nature Biotechnology*, 20:287–294, 2002.

[190] C.H. Schilling, D. Letscher, and B.O. Palsson. Theory for the systemic definition of metabolic pathways and their use in interpreting metabolic function from a pathway-oriented perspective. *Journal of Theoretical Biology*, 203:229–248, 2000.

[191] C.H. Schilling. Personal communication.

[192] C.H. Schilling, M.W. Covert, I. Famili, G.M. Church, J.S. Edwards, and B.O. Palsson. Genome-scale metabolic models of less-characterized organisms: A

case study for *Helicobacter pylori*. *Journal of Bacteriology*, 184:4582–4593, 2002.

[193] C.H. Schilling, S. Schuster, B.O. Palsson, and R. Heinrich. Metabolic pathway analysis: Basic concepts and scientific applications in the post-genomic era. *Biotechnology Progress*, 15:296–303, 1999.

[194] C. Schindler. Cytokines and JAK-STAT signaling. *Experimental Cell Research*, 253:7–14, 1999.

[195] K. Schmidt, M. Carlsen, J. Nielsen, and J. Villadsen. Modeling isotopomer distributions in biochemical networks using isotopomer mapping matrices. *Biotechnology and Bioengineering*, 55:831–840, 1997.

[196] K. Schmidt, J. Nielsen, and J. Villadsen. Quantitative analysis of metabolic fluxes in *Escherichia coli*, using two-dimensional NMR spectroscopy and complete isotopomer models. *Journal of Biotechnology*, 71:175–189, 1999.

[197] B. Schoeberl, C. Eichler-Jonsson, E.D. Gilles, and G. Muller. Computational modeling of the dynamics of the MAP kinase cascade activated by surface and internalized EGF receptors. *Nature Biotechnology*, 20:370–375, 2002.

[198] M.L. Schuler. Functional genomics: An opportunity for bioengineers. *Biotechnology Progress*, 15:287, 1999.

[199] S. Schuster and C. Hilgetag. On elementary flux modes in biochemical reaction systems at steady state. *Journal of Biologicle Systems*, 2:165–182, 1994.

[200] S. Schuster, S. Klamt, W. Weckwerth, F. Moldenhauer, and T. Pfeiffer. Use of network analysis of metabolic systems in bioengineering. *Bioprocess and Biosystems Engineering*, 24:363–372, 2002.

[201] M.A. Schwartz and V. Baron. Interactions between mitogenic stimuli, or, a thousand and one connections. *Current Opinion in Cell Biology*, 11:197–202, 1999.

[202] E. Segal, M. Shapira, A. Regev, et al. Module networks: Identifying regulatory modules and their condition- specific regulators from gene expression data. *Nature Genetics*, 34:166–176, 2003.

[203] D. Segre, D. Vitkup, and G.M. Church. Analysis of optimality in natural and perturbed metabolic networks. *Proceedings of the National Academy of Sciences of the United States of America*, 99:15112–15117, 2002.

[204] A. Seressiotis and J.E. Bailey. Mps: An artificially intelligent software system for the analysis and synthesis of metabolic pathways. *Biotechnology and Bioengineering*, 31:587–602, 1988.

[205] A.E. Setty, A.E. Mayo, M.G. Surette, and U. Alon. Detailed map of a *cis*-regulatory input function. *Proceedings of the National Academy of Sciences of the United States of America*, 100:7702–7707, 2003.

[206] A.F. Shamji, P. Nghiem, and S.L. Schreiber. Integration of growth factor and nutrient signaling: Implications for cancer biology. *Molecular Cell*, 12:271–280, 2003.

[207] S.S. Shen-Orr, R. Milo, S. Mangan, and U. Alon. Network motifs in the transcriptional regulation network of *Escherichia coli*. *Nature Genetics*, 31:64–68, 2002.

[208] Y. Shi and Y. Shi. Metabolic enzymes and coenzymes in transcription—a direct link between metabolism and transcription? *Trends in Genetics*, 20:445–452, 2004.

[209] S.Y. Shvartsman, H.S. Wiley, W.M. Deen, and D.A. Lauffenburger. Spatial range of autocrine signaling: Modeling and computational analysis. *Biophysical Journal*, 81:1854–1867, 2001.

[210] E. Sober. *The Nature of Selection: Evolutionary Theory in Philosophical Focus*. MIT Press, Cambridge, MA, 1984.

[211] A.L. Sonenshein, J.A. Hoch, and R. Losick et al. (eds). *Bacillus subtilis and its Closest Relatives: From Genes to Cells*. ASM Press, Washington, DC, 2002.

[212] V. Spirin and L.A. Mirny. Protein complexes and functional modules in molecular networks. *Proceedings of the National Academy of Sciences of the United States of America*, 100:12123–12128, 2003.

[213] S.R. Neves and R. Iyengar. Modeling of signaling networks. *Bioessays*, 12:1110–1117, 2002.

[214] I. Stagljar. Finding partners: Emerging protein interaction technologies applied to signaling networks. *Science STKE*, pe56, 2003.

[215] W.C. Stanley, G.D. Lopaschuk, and J.G. McCormack. Regulation of energy substrate metabolism in the diabetic heart. *Cardiovascular Research*, 34:25–33, 1997.

[216] M Stitt and A.R. Fernie. From measurements of metabolites to metabolomics: An 'on the fly' perspective illustrated by recent studies of carbon–nitrogen interactions. *Current Opinion in Biotechnology*, 14:136–144, 2003.

[217] G. Strang. *Linear Algebra and its Applications,* 3rd Edition. Harcourt Brace, San Diego, 1988.

[218] L. Stryer. *Biochemistry*. W.H. Freeman, New York, 1997.

[219] H. Taegtmeyer, P. McNulty, and M.E. Young. Adaptation and maladaptation of the heart in diabetes. Part I. General concepts. *Circulation*, 105:1727–1733, 2002.

[220] R. Tanaka. Scale-rich metabolic networks. *Physical Review Letters*, 94:168101–168104, 2005.

[221] I. Thiele, T.O. Vo, N.D. Price and B.O. Palsson. An expanded metabolic reconstruction of *Helicobacter pylori* (iit341 gsm/gpr): An *in silico* genome-scale characterization of single and double deletion mutants. *Journal of Bacteriology*, 187:5818–5830, 2005.

[222] I. Thiele, N.D. Price, T.D. Vo, and B.O. Palsson. Candidate metabolic network states in human mitochondria: Impact of diabetes, ischemia, and diet. *Journal of Biological Chemistry*, 280:11683–11695, 2005.

[223] J.J. Tyson, K. Chen, and B. Novak. Sniffers, buzzers, toggles and blinkers: Dynamics of regulatory and signaling pathways in the cell. *Current Opinion in Cell Biology*, 15:221–231, 2003.

[224] J.J. Tyson, B. Novak, G.M. Odell, K. Chen, and C.D. Thron. Chemical kinetic theory: Understanding cell cycle regulation. *Trends in Biochemical Sciences*, 21:89–96, 1996.

[225] H.E. Umbarger and B. Brown. Threonine deamination in *Escherichia coli*. II. Evidence for two L-threonine deaminases. *Journal of Bacteriology*, 73:105–112, 1957.

[226] S.J. van Dien and M.E. Lidstrom. Stoichiometric model for evaluating the metabolic capabilities of the facultative methylotroph *Methylobacterium extorquens* AM1, with application to reconstruction of C(3) and C(4). *Biotechnology and Bioengineering*, 78:296–312, 2002.

[227] F. van Drogen, V.M. Stucke, G. Jorritsma, and M. Peter. MAP kinase dynamics in response to pheromones in budding yeast. *Nature Cell Biology*, 3:1051–1059, 2001.

[228] A.J. Vander, J.H. Sherman, and D. Luciano. *Human physiology: The mechanisms of body function*. WCB McGraw-Hill, Boston, 1998.

[229] A. Varma. *Flux Balance Analysis of Escherichia coli Metabolism*. PhD thesis, The University of Michigan, 1994.

[230] A. Varma, B.W. Boesch, and B.O. Palsson. Biochemical production capabilities of *Escherichia coli*. *Biotechnology and Bioengineering*, 42:59–73, 1993.

[231] A. Varma, B.W. Boesch, and B.O. Palsson. Stoichiometric interpretation of *Escherichia coli* glucose catabolism under various oxygenation rates. *Applied and Environmental Microbiology*, 59:2465–2473, 1993.

[232] A. Varma and B.O. Palsson. Metabolic capabilities of *Escherichia coli*. I. Synthesis of biosynthetic precursors and cofactors. *Journal of Theoretical Biology*, 165:477–502, 1993.

[233] A. Varma and B.O. Palsson. Metabolic flux balancing – basic concepts, scientific and practical use. *Biotechnology*, 12:994–998, 1994.

[234] A. Varma and B.O. Palsson. Predictions for oxygen supply control to enhance population stability of engineered production strains. *Biotechnology and Bioengineering*, 43:275–285, 1994.

[235] D.K. Vassilatis, J.G. Hohmann, et al. The g protein-coupled receptor repertoires of human and mouse. *Proceedings of the National Academy of Sciences of the United States of America*, 100:4903–4908, 2003.

[236] J.C. Venter, M.D. Adams, E.W. Myers, et al. The sequence of the human genome. *Science*, 291:1304–1351, 2001.

[237] P.H. Viollier, M. Thanbichler, P.T. McGrath, et al. Rapid and sequential movement of individual chromosomal loci to specific subcellular locations during bacterial DNA replication. *Proceedings of the National Academy of Sciences of the United States of America*, 101:9257–9262, 2004.

[238] T.D. Vo, H.J. Greenberg, and B.O. Palsson. Reconstruction and functional characterization of the human mitochondrial metabolic network based on proteomic and biochemical data. *Journal of Biological Chemistry*, 279:39532–39540, 2004.

[239] A. Wagner and D.A. Fell. The small world inside large metabolic networks. *Proceedings of the Royal Society of London*, series B., 268:1803–1810, 2001.

[240] A.J. Walhout, J. Reboul, O. Shtanko, et al. Integrating interactome, phenome, and transcriptome mapping data for the *C.elegans* germline. *Current Biology*, 12:1952–1958, 2002.

[241] W. Wang, J.M. Cherry, D. Botstein, and H. Li. A systematic approach to re-
       constructing transcription networks in *Saccharomyces cerevisiae*. *Proceed-
       ings of the National Academy of Sciences of the United States of America*,
       99:16893–16898, 2002.

[242] P.B. Weisz. Diffusion and chemical transformation. *Science*, 179:433–440,
       1973.

[243] S.J. Wiback, I. Famili, H.J. Greenberg, and B.O. Palsson. Monte Carlo sam-
       pling can be used to determine the size and shape of the steady-state flux
       space. *Journal of Theoretical Biology*, 228:437–447, 2004.

[244] S.J. Wiback, R. Mahadevan, and B.O. Palsson. Reconstructing metabolic flux
       vectors from extreme pathways: Defining the α-spectrum. *Journal of Theo-
       retical Biology*, 24:313–324, 2003.

[245] S.J. Wiback, R. Mahadevan, and B.O. Palsson. Using metabolic flux data to
       further constrain the metabolic solution space and predict internal flux pat-
       terns: The *Escherichia coli* α-spectrum. *Biotechnology and Bioengineering*,
       86:317–331, 2004.

[246] S.J. Wiback and B.O. Palsson. Extreme pathway analysis of human red blood
       cell metabolism. *Biophysical Journal*, 83:808–818, 2002.

[247] W. Wiechert and A.A. de Graaf. Biodirectional reaction steps in metabolic
       networks. I. Modeling and simulation of carbon isotope labeling experi-
       ments. *Biotechnology and Bioengineering*, 55:101–117, 1997.

[248] H.S. Wiley, S.Y. Shvartsman, and O.A. Lauffenbureer. Computational mod-
       eling of the EGF-receptor system: A paradigm for systems biology. *Trends in
       Cell Biology*, 13:43–50, 2003.

[249] H. Williams (ed). *Model Building in Mathematical Programming*. John Wiley
       and Sons, New York, 1999.

[250] W.C. Winkler, A. Nahvi, A. Roth, J.A. Collins, and R.R. Breaker. Control
       of gene expression by a natural metabolite-responsive ribozyme. *Nature*,
       428:281–286, 2004.

[251] D.M. Wolf and A.P. Arkin. Motifs, modules and games in bacteria. *Current
       Opinion in Microbiology*, 6:125–134, 2003.

[252] M.S. Wong, R.M. Raab, I. Riboutsos, G.N. Stephanopoulos, and J.K. Kelleher
       et al. Metabolic and transcriptional patterns accompanying glutamine de-
       pletion and repletion in mouse hepatoma cells: A model for physiological
       regulatory networks. *Physiological Genomics*, 16:247–255, 2003.

[253] B.E. Wright and G.L. Gustafson. Expansion of the kinetic model of dif-
       ferentiation in *Dictyostelium discoideum*. *Journal of Biological Chemistry*,
       247:7875–7884, 1972.

[254] B.E. Wright, A. Tai, and K.A. Killick. Fourth expansion and glucose perturba-
       tion of the *Dictyostelium* kinetic model. *European Journal of Biochemistry*,
       74:217–225, 1977.

[255] B.E. Wright. The use of kinetic models to analyze differentiation. *Behavioral
       Science*, 15:37–45, 1970.

[256] E. Yang, E. vanRimwegan, M. Zavolan, et al. Decay rates of human mRNAs: Correlation with functional characteristics and sequence attributes. *Genome Research*, 13:1863–1872, 2003.

[257] R.A. Yates and A.B. Pardee. Control by uracil of formation of enzymes required for orotate synthesis. *Journal of Biological Chemistry*, 227:677–692, 1957.

[258] Y. Yokobayashi, R. Weiss, and F.H. Arnold. Directed evolution of a genetic circuit. *Proceedings of the National Academy of Sciences of the United States of America*, 99:16587–16591, 2002.

[259] L. Zheng, J. Liv, S. Batalov, et al. An approach to genomewide screens of expressed small interfering RNAs in mammalian cells. *Proceedings of the National Academy of Sciences of the United States of America*, 101:135–140, 2004.

[260] H. Zhu and M. Snyder. 'Omic' approaches for unraveling signaling networks. *Current Opinion in Cell Biology*, 14:173–179, 2002.

[261] J. Zhu and M.Q. Zhang. SCPD: A promoter database of the yeast *Saccharomyces cerevisiae*. *Bioinformatics*, 15:607–611, 1999.

Index

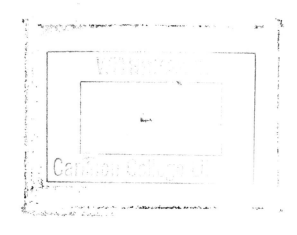